David Newell

FURROWS TURNED

This book is essentially an autobiography, but it traces in particular the changes which occurred in British agriculture during the second half of the twentieth century. The author writes from his experience of being brought up on a farm, studying for an agricultural degree, farming in a family partnership and a career in agricultural consultancy. He details consultancy experience gained in South Yorkshire in an area of small Pennine family farms which were mainly grassland, in Warwickshire where the farming was more mixed, and in Essex where the farms were large with the accent on arable crops. The consultancy section concludes with experience in management in Somerset and Gloucestershire. Later chapters deal with early retirement, travelling and a further career as a part-time Anglican minister.

DEDICATION

To my dear wife, Jean.

FURROWS TURNED

David Newell

ARTHUR H. STOCKWELL LTD
Torrs Park, Ilfracombe, Devon, EX34 8BA
Established 1898
www.ahstockwell.co.uk

© David Newell, 2017
First published in Great Britain, 2017

The moral rights of the author have been asserted.

All rights reserved.
No part of this publication may be reproduced
or transmitted in any form or by any means,
electronic or mechanical, including photocopy,
recording, or any information storage and
retrieval system, without permission
in writing from the copyright holder.

British Library Cataloguing-in-Publication Data.
A catalogue record for this book is available
from the British Library.

ISBN 978-0-7223-4739-3
Printed in Great Britain by
Arthur H. Stockwell Ltd
Torrs Park   Ilfracombe
Devon EX34 8BA

BOYHOOD

How far back can we remember? Our early memories can be tinged by what we have heard as well as what we have experienced. I can vaguely remember living at Glatton in Huntingdonshire. It was a small pretty village whose claim to fame at the time was that it was the home of the writer Beverley Nichols. He lived in a thatched cottage there. We lived in a bungalow with a slate roof. It was on a site which was slightly raised up from road level. This was just as well because a stream flowed along the roadside and crossed under a culvert outside the property. After heavy rainfall the road would flood because the water could not flow through the culvert fast enough. That culvert is associated with one of my earliest memories.

I had a twin brother, Brian, and I guess we were about three at the time. Although we were twins we were very different and that difference became accentuated as the years went by. Brian was fair; I was dark. As a baby I understand that Brian was a whiner and he suffered from a tummy rupture which may of course have been connected with the whining. I had no such ailments and was much more accommodating and happy. But, different though we were, we were always playmates, although fairly early on I became the leader and Brian the follower.

Anyway, that early memory. We were playing outside near the stream and had been joined by another boy, who must have been a near neighbour. He had discovered that if we threw a floating object into the stream, say a piece of wood, we could follow its course from one side of the culvert to the other. So we threw pieces of wood into the water on one side of the culvert, then quickly raced to the other side to see the wood reappear. Our friend had a new word for the emerging floater – it was a 'bugger'. We had not heard the word before so it was an addition to our vocabulary. Imagine the shock for our God-fearing mother when we told her that we had been playing 'buggers'. She nearly had a fit. She explained that it was a very nasty and naughty word and we were never to use it again. Our new friend was banished as a playmate.

Looking back it is amazing what three-year-olds did in those days and how trustful parents were. My father was a farmer. He had started farming on his own account when he got married in the mid 1930s. Previous to that he had worked on his father's rented farm at Middlemarsh Farm, Sawtry, just one mile distant from Glatton. The farm at Glatton was called Butts Farm. The farm buildings and land were on the opposite side of the road to the bungalow. The original farmhouse must have been separated from the farm sometime earlier, possibly associated with a retirement. The farm was about 150 acres of boulder clay soil, together with farm buildings for cows and storage. The land was cropped with cereals and grass for the cows and followers. At the age of three and upwards my brother and I freely roamed around the buildings and the land without any supervision or even anyone knowing where we were at a particular point in time. We had to cross a road, of course, to reach the buildings, and the land, but there was little traffic in those days. My parents did possess a small car, a blue Austin 10, which was invaluable for business and also for my mother to visit her mother, who lived in Huntingdon, some ten miles away.

Mentioning that my parents did not always know where we were evokes another early memory. The farm buildings included a cowshed and also an enclosed yard for the cattle. Somehow, one day I became trapped in the enclosed yard. My brother was not with me at the time. I am sure I must have shouted, but no one came to my rescue. I could not open the large solid door out of the yard, so using my ingenuity I crawled underneath it. But it was a cow yard and between the door and the floor was cow manure. Imagine what I looked like and smelt like after I had crossed the road home! Not a pretty sight!

On another occasion my father had hired a bulldozer to level some rough ground on the farm. The bulldozer duly arrived attached to a massive D6 Caterpillar tractor. It was carried on a heavy-duty low-loader. I cannot remember where my father was. Perhaps he had gone on ahead to open the gates. Anyway, I was appointed to show the driver where the field was. Imagine it – a child of three guiding this massive piece of equipment to the right field. It was such a thrill, which is probably why I remember it so well. Nowadays, a child this age would not be allowed anywhere near such machinery yet alone ride on it. You are not allowed on a tractor at all unless you are at least twelve.

So there are three early memories of life at Glatton. I cannot remember much more except the geography of the bungalow and looking out of a bay window over the garden. Also our next-door neighbours lived in a thatched cottage. They were Mr and Mrs Saunders and he was bedridden. He had been a farmworker, but had fallen off a corn stack and broken his back. I guess he was no more than middle-aged at the time. He was destined to spend the rest of his life in bed. No treatment then for a broken back, no physiotherapy and no compensation either. I don't know how they managed to live. Mrs Saunders used to take in washing as just one meagre source of income. I never remember Mr Saunders going outside again. I guess he was too heavy either to be manoeuvred into a wheelchair or for Mrs Saunders to push him even if he had been lifted into a chair. It is likely that they would not have been able to afford a wheelchair even if they could have used it. No state health service or social services to provide in those days.

Times were changing though. War had been declared on Germany. Farming was becoming more important in the national interest! We depended on imports of food to keep the nation alive and imports were severely threatened owing to the action of German U-boats. As a nation we needed to produce more of our own food. In the 1930s farming was at a low ebb as the country depended on cheap imports to keep the nation fed. Scrimping a living from a simple family farm was a meagre existence. But now increased home food production was vital if the nation was not going to be starved into submission.

These changes coincided with a change in my family's circumstances also. My father's rented farm at Glatton was sold to a new landlord, who happened to be the local rector. The era when vicars and rectors belonged to the landed gentry was not quite extinct. This rector decided he wanted to farm the farm himself, so my father's tenancy was terminated. There was no security of tenure then. That was to come later with the 1948 Agriculture Act.

Fortunately for my father, my grandfather decided to retire from farming at Middlemarsh Farm. The tenancy was offered jointly to my father and his brother. Conveniently for them there were two houses on the farm. There was an older period house down a drive from the road adjacent to the farm buildings. Standing just back from the road was a newer red-brick house built in the early twentieth century. This was to be our new home. I cannot remember anything about moving, but I can remember my grandfather's sale as he retired. I remember all the implements standing out in the field ready for the auction. My father of course brought his own machinery tackle from his previous farm, and anything else the brothers wanted they had to buy in the sale. It would be over sixty years before there was another Newell sale at Middlemarsh.

Fairly soon after we arrived at Middlemarsh my brother and I started school at Sawtry County Primary School. Our birthday was in February and we were able to start at the beginning of the term when we were five. It was over a mile to school so we had to be escorted there each day. My mother employed a helper in the farmhouse who joined her straight from Sawtry School at the age of fourteen. Her name was Betty and she eventually became like an elder sister. She accompanied us to school with her bicycle. The bicycle was allowed so that she could ride back home to the farm. However, as soon as she was out of sight of the farm on the way to school she would ride us on the bike, one on the carrier and one on the seat. So much for health-and-safety regulations! We would have to alight of course when we reached the village. Villagers have ears and eyes and mouths!

School was quite a family affair. A close friend of my mother's was the infant teacher and my mother's sister also taught in the school at a higher age level. Very soon my mother was recruited also, owing to the shortage of teachers during the war. She had been trained at Homerton College, Cambridge, and was an able and respected teacher. Having these family members and friends on the teaching staff caused a few difficulties for my brother and me in relationships. I suppose the other children were always suspicious of what we might pass on. I was earmarked as a bright child early on, and I even skipped several classes later in my primary school career. By the time I was ten I was in the top form. There most pupils had to stay until they were fourteen unless they passed the eleven-plus and went on to grammar school. The rate of eleven-plus passes was no more than two or three per year from between twenty and thirty children in the class taking the exam. I was expected to pass the eleven-plus; Brian was not. This caused considerable anxiety to my mother, who felt that as twins we should be treated alike. Brian was coaxed to obtain a higher performance level and reach my standards. It was impossible for him to do this and he should have received more sympathetic parental understanding. Instead the way he was treated led to him developing an acute sense of inferiority which has persisted throughout his life. It amazes me now that an able teacher like my mother did not possess better psychological understanding.

Our life at home revolved very much around the annual cycle of the farm year. Sowing the cereal crops and harvesting the grain were the highlights of the farming season. The Ministry of Agriculture took increasing control of what crops were grown and the standard of husbandry achieved. We had to grow some flax for linen production and also sugar beet for sugar. Orders were given to plough out more grassland for cropping. We had two fields called the 'rough fields', which is self-explanatory. They were fairly narrow 'ridge and furrow' in contour, inherited from farming systems practised decades earlier. The grassland was coarse, and in need of improvement. To cope with the initial cultivation my father hired a Gyrotiller. This was like a very large tractor with revolving cultivator blades on the back. It was an awesome machine shredding tree roots around the headland of the field as though they were pieces of string. It certainly prepared the fields for more refined cultivation, but a row of lovely elms along one side of the field died subsequently. I suppose the disturbance of the roots was just too drastic for the trees to recover from.

During the war harvesting cereals was a major operation. First the crop was cut with a binder which converted the field of corn into rows of sheaves. The binder mechanism was driven from the main land wheel through a series of chains. When the going was easy and the crop standing this was fine. But if the cereal crop had been battered by the elements, and was lying rather than standing, or if ground conditions were wet, the main ground wheel would slip or slide. The mechanism then came to a standstill, causing much frustration all round. Another source of frustration was when the binder missed tying the sheaves, which happened often. The tying mechanism, known as the knotter, was a very temperamental part of the equipment. A few years later new binders were powered by a power take-off shaft driven from the tractor. This was a great improvement as the binder mechanism kept going even in rough conditions. The knotter mechanism, however, was not perfected. It still missed tying some sheaves.

The binder went round and round the field throwing out the sheaves in precise military rows. As cutting the corn neared the centre of the field there was usually great excitement. Rabbits and other vermin, like rats, were driven to the centre as the binder went round and round. They were seeking refuge in the only shelter left for them. But eventually, taking fright, they had to run for it. By that time farmworkers and boys like myself had surrounded the central strip armed with sticks. The majority of rabbits did not escape because they had no sense of direction in a landscape which had completely changed from a few hours earlier. A few ran along by the rows of sheaves where they had an uninterrupted run, and they usually escaped unless they were shot. Most tried to run through the rows of sheaves, which was slower and much more difficult. They often bumped into sheaves as they tried to find a way of escape. That escape was usually terminated by a heavy blow from someone's stick. I suppose it sounds a bit barbaric now, but we regarded it as great fun and a bit of sport amidst hard work. Rabbits were prized then and sold well in the local markets. Rabbit meat was prized in the countryside during the war. It added to the meagre diet and provided extra animal protein when other meat was rationed. It was before myxomatosis decimated the wild rabbit population and effectively killed off the market for wild rabbits. It is not surprising that myxomatosis stopped the demand for rabbit meat. A rabbit infected with the disease is a gruesome sight. But it was a blessing in disguise. Rabbits had become a major pest on farms and throughout the countryside there must have been millions of them. Their burrow warrens were sited along hedgerows and in woodland. From there they would graze out into field margins and do considerable damage. Cereal yields were significantly lower where rabbits had grazed and over the country as a whole that amounted to many tons of grain.

    After cutting, the sheaves were gathered up by hand and stooked. This operation meant arranging the sheaves in a V-shape propped up against each other. Each stook was composed of eight or so sheaves. The field of sheaves was thus converted to a field of stooks, which were left for a few days to dry out further before they were carted to the stackyard. Sometimes stooking was a painful process, especially if thistles were embedded in the sheaves. It was before sprays were widely used to kill weeds.

    The year when I was eight I drove the tractor for all of the harvest period. The tractor was a Fordson Standard, the workhorse of farming during the war years. It was a simple piece of machinery by today's standards powered by a paraffin engine which had to be started with petrol until the engine warmed up. I drove this machine with the binder and later pulling the trailers hauling the sheaves to the stackyard. Three trailers were employed for this: one in the field being loaded, one at the stackyard being emptied, and the other either being ferried back to the field or en route to the stackyard. Three people were required in the field for loading, at least three for unloading and stacking at the stackyard and then one of course driving the tractor to and fro.

Fancy a boy of eight being trusted and able to cope with the tractor driving. As I mentioned earlier no one is allowed on a tractor now until they are at least twelve. However, tractors are bigger and more powerful today than they were then. But I was cheap labour – I saved the wages of a man. However, I was paid a small amount. My brother was not permitted to drive the tractor yet. My uncle quipped that he was not old enough. His job was to lead the horse pulling the trailer from stook to stook in the field. I found out later that he was paid slightly more than me because he had a more tedious job. I would still rather have driven the tractor!

I was too young to remember much about the war. At Sawtry we were surrounded by new airfields as the aerial bombardment of Germany intensified with the advent of America into the war. I can remember lying in bed at night and hearing the droning of aeroplanes as they took off or returned from sorties over Germany. One night my brother and I were awakened by a frightened mother. It seemed like daylight as incendiary bombs rained down from all around. Whether the Germans were emptying their bomb loads before returning home or were off target I don't know. The cornfields were nearly ripe and each bomb burnt a large circle around it. Fortunately the corn was not ripe enough for the whole field to be set alight. We all sheltered under the stairs in the farmhouse for safety. Fortunately no bomb dropped near enough to damage the houses or other buildings. The next morning we discovered a German Dornier had crashed in a neighbour's field, leaving a huge crater. A dead airman was found in a field nearby. Presumably he had baled out of the crashed plane and his parachute hadn't opened in time. The area was soon cordoned off. That really was my only experience of the war except the daily news reports and the constant nightly drones of the aeroplanes.

As the war progressed increasing numbers of enemy prisoners arrived in Britain. A prisoner-of-war camp was established in Sawtry. The prisoners were escorted to farms to provide extra labour. I remember us employing Italians, later Germans and also some Austrians. The Italians were fun and sang a lot. The Germans worked hard. The Austrians were surly and idle. A Women's Land Army hostel was also established in Sawtry and we employed land girls from time to time. I remember one land girl was sent to roll a field with a tractor and roller. She was told that someone would wave a flag from the farmhouse when it was lunchtime. When the flag was waved she promptly stopped the tractor in the middle of the field instead of proceeding to the headland. That caused much amusement, to her embarrassment. Also I remember my uncle, who was single, choosing one of the land girls as his girlfriend.

Shortly after the war we had a very severe winter. It was 1947. I cannot remember whether the cold weather started before Christmas, but I do remember that skating had become a popular pastime by the time of my birthday, which was 1 February. My father took my brother and me to skate at Cowbit Wash, near Crowland. I had a pair of skates called Fenland runners. They were mounted on a wooden base and were curled up at the front. They were secured by a screw into the heel of a boot and straps across the toe. Shortly after this, Bury Fen at Earith became a popular skating centre and the National Skating Championships were held there. As a family we went to Bury Fen skating a number of times that winter – including my grandfather, who was seventy. He was a wonderful skater, gliding along gracefully on his long Fenland skates with his hands behind his back. One day, a would-be skater offered him £10 for his skates. Skates had become so scarce because of the popularity of skating that it was impossible to buy a pair new. My grandfather took his skates off and sold them. I guess it was the easiest £10 he ever received, and he said at the time that he was too old to think of skating again another season.

By March, the whole country was paralysed by blizzards and snowdrifts. Our farm was completely cut off, the snow lying level with the tops of the hedges on each side of the road. My mother tried to walk to school one day, but abandoned the attempt when she sank in snow up to her neck. The farm was cut off for a whole week. I suppose the milk from the cows must have been thrown away because it could not be collected. Eventually the council managed to get a snowplough to Sawtry. My father and others managed to haul the plough from Sawtry to Glatton with an R2 Caterpillar tractor. This opened up communications again with just a single-track lane connecting the two villages. When all the snow eventually melted it caused chaos in the low-lying Fenland of East Anglia. The waterways just could not cope with the volume of water flowing in from the surrounding higher land. There was serious flooding.

My parents were staunch Methodists and the family attended Sawtry Methodist Church every Sunday. We usually attended the six o'clock service. Morning services were not so popular then. My father was a strict Sabbatarian and used to read a Lord's Day Observance Society tract called *Joy and Light*. I can honestly say that there was no joy and light in his Sunday observance. It was a case of the law before the spirit. He often got into nasty arguments with ministers and local preachers if they prescribed a laxer approach to Sunday activities. My mother was more liberal, but she kept quiet to keep the peace. She was often embarrassed by these arguments, but he was insensitive to her feelings because he was always right! She did not qualify as a Methodist Local Preacher, but she did become a helper on the preaching plan. This meant taking occasional services in the Hunts Mission chapels. My father always drove her to these appointments, even though she was quite capable of driving herself. Was this his way of controlling what she said in her sermons?

BOARDING SCHOOL

As my brother and I reached the age of ten my mother became increasingly anxious about our future education. She had become resigned to the fact that Brian would not pass the eleven-plus in spite of her efforts, which probably were a hindrance rather than a help. In any case the great majority of children did not pass the eleven-plus, so Brian was no different from them. It did not seem to be appreciated at the time that I was the exception rather than the norm.

My mother made enquiries about sending us to boarding school. I think Oundle, Stamford and Kimbolton were considered. Eventually Kimbolton was selected because the fees were lower. We went there to take an entrance examination and met the headmaster, Cyril Lewis. He had only been at the school for a term, having succeeded the previous headmaster, William Ingram. We both passed but were placed in different forms, Brian in 1b, me in 2a. So from the beginning we were placed in different forms, as we had been at Sawtry School. However, my mother's objectives had been achieved. We were placed in the same school, so had equal education opportunities even though we were clearly unequal in ability.

We started school at Kimbolton in January 1948. What a traumatic experience! We had not been prepared psychologically for boarding school as many children are from an early age. It had all been arranged very hurriedly. On the farm and at Sawtry School we had enjoyed a comparatively free life. Village school life was easy, and I was virtually top dog even though I was only ten. Out of school we were free to roam around the 280 acres of the farm, bird's-nesting or doing whatever was appropriate at the time of year.

Imagine the privation of a boarding school, herded into a dormitory with twenty other boys. The dormitory was in the main school block, looked after by a master called Padley. He became the headmaster of the preparatory school when that was established a few years later. There were two other twins in the dormitory who were identical, as distinct from Brian and myself. School life followed a strict routine every day of lessons and after-school prep. Sport was important too, of course, so that became a welcome relief. We were allowed two outings per term with our parents, on either a Saturday or a Sunday. I think this was limited to ensure that boys did not suffer too much from homesickness. I remember that I floundered for the first two terms at least. I also suffered from a succession of boils, which I expect were psychosomatic in origin. From being top dog at Sawtry School I became a little fish in a big pool, almost anonymous. I had to cope with additional subjects like French and Latin, and the other boys in the class had a term's start on me, so I struggled. Form positions were posted on the noticeboard monthly. My position was lost somewhere in the middle, about fifteenth in a class of nearly thirty. I did take the eleven-plus exam when I was eleven and passed. Fortunately for my parents Huntingdonshire had a scheme for paying the tuition fees at Kimbolton for those who passed the exam. This was because Kimbolton was sited in the county. The only other grammar schools in the county at the time were Huntingdon, Ramsey and Fletton. So me passing the eleven-plus saved my parents a great deal of money.

After two terms at school it was the summer holidays. The boarders at Kimbolton used to anticipate holidays by singing a school ditty. I cannot remember all of it, but the following is lodged in my memory!

> No more stale bread and butter
> No more cocoa from the gutter
> No more beetles in my tea
> Making goggly eyes at me.
> No more spiders in the bath
> Trying hard to make me laugh.
> When the train goes chuff chuff chuff
> I'll be on it sure enough.

Those eight weeks of holiday passed far too quickly, so it was soon back to school. The situation was changing at Kimbolton. The school had bought Kimbolton Castle, together with its grounds. This was the ancient seat of the Earl of Manchester, the place where Henry VIII's first wife, Catherine of Aragon, was banished to. It was a considerable acquisition for the school, but it demanded enormous investment before it was habitable as a school. The castle had been neglected and unlived in for a number of years. However, the surrounding woods and parkland were magnificent. The acquisition improved considerably the whole environmental aspect of the school. Whether the ballroom and panelled hall in the castle were ever right for a boys' boarding school is perhaps more debatable. But it did ensure the future of the castle as a national monument and provide the finance for its upkeep.

Returning to school after that first summer holiday, Brian and I moved accommodation to Kimbolton House in the High Street of the village. The school owned a number of houses which were used for boarders and also provided accommodation for housemasters. Kimbolton House was quite large, housing about fifty boarders in dormitories. From Kimbolton House we had to march in a crocodile form to and from the main school. It was an early start because we had to have breakfast at the main school before school started at nine o'clock. There was a formal inspection every morning before the crocodile could set off, to make sure we were all smart enough and our shoes shone. These crocodile marches were unaccompanied except by prefects. There was a nasty accident somewhere in the country when a bus driver ploughed into a school crocodile at night. After that our school crocodiles had to have a lantern at the front and the back after dark.

The housemaster was called Gibbard. He was a lifelong bachelor with the nickname of Bant. I think this nickname originated from the fact that he had been a bantamweight boxer in his youth. He was a strict disciplinarian, extremely loyal to the school and to the previous headmaster, William Ingram. A new headmaster brought changes, of course, and I think Bant, although he was nominally deputy head, found it very difficult to adapt to these new ideas. I can still hear him saying, "In the good old days of W. Ingram . . ." Obviously he bitterly regretted the passing of those days. My brother and I were placed in the same dormitory again, albeit this time in a more intimate room of only six beds. Close friendships were formed with the other boys which persisted for the rest of my time at the school. This time of adolescence is associated with growing sexual awareness, which was often the subject of discussion at night before we drifted off into sleep.

Sport was an important part of the school's curriculum. The main sports were football in the winter and cricket in the summer. The school was divided into houses for internal competition – Dawson, Bayle and Ingram. I was placed in Ingram House, named after the previous headmaster. There was fierce competition between the houses each year to accumulate the most points and thus become the most prestigious house.

School football and cricket matches were arranged against public schools within travelling distances and various Cambridge colleges. We were encouraged to support our teams from the touchline when we were free to do so. I can remember watching many cricket matches on sunny Saturday evenings, which was great relaxation after a busy week. I was disappointed never to be selected to represent the school at either cricket or football. Many of the boys had cricketing fathers who occasionally played against the school. It always seemed to me that their sons were always first choice for teams

– a sort of old boys' network. Perhaps that is sour grapes or maybe I was just not good enough! However, before I left school I did captain my house football team and we won the house competition that year against the odds.

One of the masters, Kiffin Owen, seemed to be closely associated with the old boys' network. He was one of the old stalwarts, a friend of Bant and resistant to change. He was the history master, which was very appropriate because he lived in the past. I remember his history lessons as being utterly boring and devoid of creative thought.

Soon after I started school at Kimbolton a new sports master was appointed named Stringer. He completely revolutionised the school's approach to sports, particularly athletics. He took a much more professional and modern attitude to the whole sports scene. I remember, in my early days at Kimbolton, house points were awarded for who could change into sports gear the quickest – hardly an athletic feat! Stringer scrapped all that and points were awarded only for athletic achievement. He introduced new sports into the curriculum, like cross-country running and field events such as javelin throwing and shot-putting. Previously the only field-like event was throwing a cricket ball.

I excelled at athletics and gymnastics and represented the school in both these sports. I competed at county level in cross-country running and sprint events. One year our 4×100 yards relay team won the Beds & Hunts Championship when Brian was the lead man and I ran the final leg. I still have a medal to prove it! Mr Stringer was delighted with the result. I used to do well in the school's annual sports day, which was a prestige occasion watched by the majority of parents and other dignitaries. When I was sixteen I won the 440-yard race. Unfortunately I was unable to compete the following year, my last year at school, owing to a groin strain, which was a great disappointment.

In my second year at school, form 3a, I began to find my feet academically. I had caught the others up in French and Latin and my English and maths were above average. Maths had extended from arithmetic into algebra and geometry, which I enjoyed. Extracurricular activities were also introduced, and I was able to join the Young Farmers' Club. I remember an uncle joking yet again that he had heard David had joined the Young Farmers' Club but Brian was not old enough yet! County competitions with other clubs, such as public speaking and rallies, took us outside the school. I was selected to represent the club at a public-speaking competition held at Huntingdon. The team comprised a speaker, a chairman, a proposer of thanks and a seconder. Not surprisingly we won the competition. I say not surprisingly because we were coached by schoolmasters and were starting at a higher academic level than the other teams.

The GCE syllabus started when I reached form 4a. This coincided with a number of changes in the school routine. Kimbolton Castle had now become habitable and the main body of the school moved there for accommodation, meals and lessons. That was in 1950. The top floor of the castle was used for dormitories, the middle floor for lessons and the bottom floor for various other activities as space became habitable. There was still a lot of work to do on the ground floor. Gibbard moved from being the housemaster at Kimbolton House to housemaster at the castle. The new headmaster's study was sited on the middle floor in a room previously occupied by Catherine of Aragon. The stable block of the castle was converted into kitchens and a dining hall. The previous school buildings, sited about three-quarters of a mile from the castle were redesigned to accommodate the science laboratories and the preparatory school, which was introduced at this time with a separate headmaster. The old school and the castle were connected by a scenic drive about half a mile long through woodland. This made access between the two sites easy as well as safe.

Starting the GCE syllabus had a dramatic effect on my academic standing. Science was introduced into the curriculum at this stage, and this included physics, chemistry and biology. Within a few weeks of these subjects being introduced into the curriculum my form position rose from being about fifteenth to within the top five. I stayed within the top five for the next two years in the fourth and fifth forms. You can imagine what a boost this was to my ego after being that small fish in a large pool. What became clear to me, and also my schoolmasters, was that I was stronger on the science side than the arts. This made the choice relatively simple when it came to deciding what subjects to concentrate on for A level. However, before then there was a problem. I was a year below the average age of my form. This meant that at the time for taking the GCE O-level exams I was only fifteen. There was a ruling in operation at the time that these exams could not be taken by pupils until they were sixteen. So I was not allowed to take the exam. I sat all the papers, but they were marked by the appropriate schoolmaster, instead of being forwarded to the examination board. What bureaucratic nonsense and how unjust on those caught in the trap. The ruling was relaxed after a number of years, but that did not help those already affected. I faced the choice of either staying down in the fifth form for another year or else going on to the sixth without any GCEs. There were problems both ways because some of the syllabuses would have changed for GCE. For instance the English-literature material would have changed because it was a two-year syllabus. I was advised to go on to the sixth form and to take GCE Latin, French and English language at the end of my first sixth-form year. This was what I did eventually, but it carried the risk of leaving school with no GCE qualifications at all. The masters had more confidence in me than I had in myself. I duly took English language, French and Latin at the end of the first year in the sixth form and passed all three. This coincided with my brother taking all his GCE subjects in form 5b, passing in English literature and geography. He left school after that and returned home to work on the farm. I carried on at school for another year in the upper sixth form. My main subjects were biology, chemistry and physics.

Joining the sixth form opened the possibility of being a

prefect. These were selected by the headmaster. One first became a sub prefect, which was a sort of probationary period. There was little difference in responsibility between sub prefects and prefects, except the latter wore jacket-length gowns. This certainly gave them more status and it was quite a proud moment to receive a gown. I believe I received my gown at the beginning of the upper sixth form.

The sixth form at Kimbolton was fairly small in those days so we were quite an elite group. We got to know each other extremely well and formed some close friendships. There were only three of us studying for A-level biology, one of whom was Joe Norman who became a strong personal friend. He came to stay with us at Middlemarsh Farm one summer holiday to help with the harvesting. He was another of the school's prefects. Day boys tended to miss out on prefects' responsibilities because most of the duties were associated with boarders' activities.

Prefects' duties included supervising prep duties, being responsible for a table at mealtimes, helping maintain discipline and assisting masters when required. One extra duty I was given was being a laboratory steward for the physics laboratory. This involved assisting the physics master in servicing the equipment in the laboratory ready for lessons and keeping it clean and tidy. There was a small remuneration for these duties which amounted to ten shillings per term. As this was not allocated until the end of term it was useful pocket money for the holidays.

I acquired another extra duty by virtue of my ice-skating ability. There were two large ponds on the castle estate which during severe winter weather froze over and were suitable for skating. I was given the job of testing the ice to see when it was safe. No master accompanied me to confirm that the ice was safe and several boys joined me on the ice. Two boys would persist in skating round the pond together even though the ice was creaking badly. The inevitable happened and the ice broke, possibly over an inlet to the pond. Both boys were pitched into the icy water. Luckily we got them out, unscathed apart from shivering violently. I sent them off for a hot bath and no one was any the wiser regarding their accident. I advised the headmaster that the ice was creaking and I did not think it would be safe enough for another couple of days!

Another prefect's duty was supervising the boys in the various boarding houses. I was allocated to White House at the top of the High Street, opposite the village church which all the boarders attended for matins on Sunday. The housemaster at White House was Tom Pierce, who had a fearsome reputation as a disciplinarian. When he was on duty at mealtimes at the school he used to stand up to make any announcements. As soon as he stood up a hush would commence at the top table end and proceed like a wave down the whole of the dining hall. Other masters used to have to shout before they could get a hearing. However, I discovered that underneath that surly exterior Tom had a soft heart and I became one of his favourite pupils. I guess that he requested me as one of his prefects at White House and also as his lab steward in the physics laboratory.

White House accommodated about twenty boarders, and there were two prefects. The prefects' room was dire because it was situated between the dormitories and the washrooms. All the boys had to pass through this room on their way to the toilet or the bathroom. It was good in that the prefects always knew what was going on. But it was poor accommodation for pupils studying for A levels, and it was always cold because it was at first-floor level over the garage. Tom Pierce showed some consideration, sometimes inviting us down to his lounge for a sherry and to watch television, which was very much a novelty then.

During my time in the upper sixth I began to start thinking more seriously about my future. I was advised that my academic attainment in my chosen subjects was good enough to consider going to university. My strongest subject was biology and I was deeply attached to farming as a career. I decided to apply for a university place studying agriculture. As I was only seventeen I expected that I would have to take a year out even if I qualified for university entrance. Only about two per cent of scholars qualified for university then, as opposed to the fifty to sixty per cent now. I applied to both Reading and Nottingham. To my surprise I was accepted at Nottingham straight away provided I passed my three A levels. I was entered for A-level physics and chemistry and scholarship-level biology. I managed to pass all three and another bonus was to be awarded the school biology prize. So I was on my way to the University of Nottingham, but not until I had spent another eight weeks helping with the harvesting at Middlemarsh Farm.

UNIVERSITY

I went up to the University of Nottingham in September 1954 after harvesting on the farm was complete and the autumn cultivations were well under way. The Agricultural Department was situated at Sutton Bonington. Before being taken over by the University the site had housed the Midland Agricultural College. This was one of the big three agricultural colleges in England, along with the Royal at Cirencester and Harper Adams in Shropshire. It had concentrated on preparing students to take the National Diploma in Agriculture (NDA) and also the National Diploma in Dairying (NDD). With the takeover by the University in 1948 the majority of courses were switched to degree status, except the NDD, which was retained for a number of years. This was a social bonus for the college campus because the NDD course brought in most of the girls!

Before term started, I attended a freshers' course at Sutton Bonington. This introduced us to the facilities available at the college and also the extra facilities of the main university in Nottingham. There was a good connection between the two sites as the railway station at Kegworth was only about half a mile from the college. The majority of students at Sutton Bonington were housed in hostels, although a few students lived out in lodgings. There were three men's hostels and one women's, so that illustrates the ratio between the sexes. It was a woman's world as far as choice went!

I was placed in Hostel 1, which consisted of mainly twin

rooms with washing facilities on each of the two floors. My room-mate for the first two years was Ernie Bathurst, who had completed his national service so was several years older than me. It is a bit of a gamble throwing two people together to share a room, but the system seemed to work quite well. Ernie and I got on famously, although we were quite different. Initially Ernie was brasher, more extrovert and more assertive than me, which reflected his wider experience. I was quieter and more prepared to feel my way into university life. But we soon learnt that we were both Christians, and that was a great starting point.

University life was very different from school. I soon learnt that it was a big step into adult life. Lecturers were not like schoolmasters, and there was no established routine. You were free to make your own decisions and establish your own order of priorities. You could wander on and off the premises as you liked. I took my bike to college, which added to the sense of freedom. A few students possessed a car or a motorbike, but it was only a handful. I had to adjust to meeting and socialising with members of the opposite sex, which was a new experience for me. Having previously been confined to the school or the family farm, members of the opposite sex were foreign territory.

My degree course in agriculture was science-based for the first two years with the opportunity of specialising in the third year. It was a new experience taking down notes from lectures and having to rely on them and textbooks to acquire the necessary knowledge. Some lecturers were well prepared, methodical and easy to listen to. It was straightforward taking down good notes from them. Other lecturers were nothing like so structured or ordered, and taking sensible notes from them was a nightmare. It usually meant a lot of extra reading to get to grips with the subject. The most difficult subjects for me were physical and organic chemistry, which require quite a high degree of mathematical skill. We had exams at the end of each term, which were a good test of whether we were mastering the subjects or not. I remember one student saying that it was like taking A levels three times a year. I worked quite hard and managed to keep up with what was required. The range of subjects included botany, zoology, microbiology, geology and soil science as well as agriculture and horticulture. A few students who either had not worked hard enough or who had failed to cope were asked to leave at the end of the first year.

As the college at Sutton Bonington was rather isolated from the main university it developed a social life of its own. Student activities were centred generally around the student common room, which provided coffee and tea and was also licensed. It had a games room, which included table tennis and was used for student dances mainly at weekends. I don't remember attending any dances during my first year. I was probably too shy, and anyway I had never had any dancing lessons. Looking back, I do not know why the students didn't organise dancing classes for the novices.

Table tennis was another matter. I soon established myself as one of the better players in the college. It was also an opportunity to improve my social skills with mixed table tennis. Fairly soon I became established in the college men's team, and I have happy memories of representing the college against teams from Nottingham, including Player's Cigarettes.

Among the groups and clubs, there was a Christian Fellowship in the college. The main Christian groups in the colleges at that time were either affiliated to the Student Christian Movement or the Scripture Union. I remember debating whether we should be affiliated to either of these groups. Eventually we decided to join neither and to remain independent as a Christian Fellowship. The fellowship organised a short service at the beginning of each day and also had a meeting on Sunday afternoons to which a speaker was invited. The meeting, which was held in one of the hostels, usually finished with us enjoying tea together. It was during my time at college that my Christian faith deepened. I was elected on to the committee of the fellowship as prayer secretary. My main responsibility was organising the daily service, which always had an invited leader. These included students, but also several of the lecturers. The chairman of the group was Brian Heap and the secretary was Stan Cramer. Both of these became personal friends and still are fifty years later.

One weekend we invited the famous China missionary Gladys Aylward to come and speak to us. Her story was told in the film *The Inn of the Sixth Happiness*. She surprised us by accepting our invitation and coming for the whole weekend, staying in the women's hostel. She held us spellbound as she talked about her missionary work in China and how she had led those Chinese children over the mountains to escape the advancing Japanese army.

I was brought up as a Methodist and there was a small Methodist church in Sutton Bonington. A group of us from the college used to go there regularly to the Sunday morning service. This brought us into contact with the superintendent minister, the Reverend Bert Hodgetts from Castle Donington. Occasionally he would come to the college and lead morning prayers. On one occasion a busload of us went over to Castle Donington to a circuit rally. It was addressed by the Reverend Dr William Sangster, who was a household name in Methodism at the time. He probably came to Castle Donington when he was President of the Methodist Conference. He was a compelling speaker and a person who obviously lived close to Christ.

Sutton Bonington had some good sports facilities and fielded football, rugby and hockey teams. I mostly played football, but was occasionally cajoled into playing rugby if they were short of players. My speed at a three-quarters position proved a valuable asset. I represented the college in the football second eleven and we had fixtures most Saturday afternoons during the winter terms. I remember one of the roughest teams we played against was the trainee priests at Kelham College.

Again, it was in athletics that I really excelled. I did some cross-country running during the winter and track events during the summer. My speed combined with my stamina led me to concentrate on the quarter mile. I represented the college in this event for the three years I was at college. The highlight of the season was the triangular athletics match between our college, the Royal Agricultural College and Harper Adams. In my final year, I won this event at the Royal and was awarded my college athletic colours. During my final year, I was encouraged to enter the university trials at Nottingham. To my surprise I made it into the team easily. I was only sorry that I had not been encouraged to attend the trials earlier. So during my last term I ran for the university most weeks in the quarter-mile and was often selected for the relay races as well.

During my second year at Sutton Bonington I began to mature as a student. During my first year I had been adapting to the change from school life to university. I was still only eighteen, so well below the average age of my fellow students. I began to feel socially more confident and started to go to some of the dances, encouraged to learn a few of the basic dance steps by some of the girls. The new intake in my second year included a number of attractive girls, which raised testosterone levels among the males. It was later in that year that I made my first real date, taking one of the girls to the cinema in Loughborough to see the film *High Society*, starring Grace Kelly, Bing Crosby and Frank Sinatra – very romantic. However, the romance did not last! I think that there was too much competition around. I did become very fond of one of the girls, but alas she had her eyes on someone else.

One highlight of the week was called 'Dining In Night'. After a meal there was some eminent person who came to speak. I guess the purpose of these nights was to widen our perspectives. I know that we had some good speakers, but I cannot remember many of them now. One who is retained in my mind was Frank Henderson, who wrote the book *The Farming Ladder*. He and his brother farmed successfully on poor land in Oxfordshire during the depression of the 1930s. He was a real countryman and a charismatic speaker. I think that he was genuinely surprised to get such a good reception from a student audience.

As the two-year science-based course came to an end we were encouraged to think about specialising in the third year. We seemed to have covered everything in those first two years from the classification of insects to the classification of plants. We had also had a course on statistics. What you did in the third year was determined by how well you fared in the exams at the end of the second year. Those with the best results and showing the most promise were invited to do an honours degree, which included doing a thesis on your specialist subject. I was not quite sure at that time whether to concentrate on animals or plants and I did not have a mentor to encourage me one way or the other. Looking back, I should have opted for plants and done a thesis on wild oats, which were becoming a serious weed in the cereal-growing areas of the country. One lecturer encouraged me to do an agricultural educational tour on the Continent or in the United States of America during the summer holidays between the second- and third-year courses. I declined because the family expected me at home to help with the harvest, as they had done since I was eight. It would have been good for my father and for me if I had opted out of harvesting that year and gone abroad, but it would not have been well received. Looking back, I can see that the family were becoming too dependent on me, and that led to repercussions later.

During my university course there had been some tension between my father and me over funding. In my first year the Huntingdonshire County Council paid my tuition fees and my father had to pay for my accommodation. He did this with rather bad grace, which left me feeling awkward to say the least. My father had not helped the situation because he had been uncooperative with the council in disclosing details of his income, on which council grants were based. After arriving at college I discovered that the majority of students were on much more generous grants than I was. This led me to seek an interview with the chairman of the Huntingdonshire County Council Education Committee. My father forbade me to go and see him, thinking no doubt there would be more questions about his level of income. My mother encouraged me to go. That was one of the few times I can actually remember my mother defying my father. I went to see the chairman anyway, and as a result I was placed on a full grant for my next two years at university. Although this saved my father quite a lot of money at the time, he never commented or thanked me.

Following the second-year exams and considering my

preferences I was placed on a pass degree course in the third year, specialising in the economics of farm management. I was still thinking about farming as a career, which was influenced by the fact that the family needed me at home. My father was only of average ability as a farmer and the farm was barely making a profit. The tutor for the course was Stuart Senior, a lecturer in the Economics Department of the university. The course proved practical, interesting and challenging. It was based on looking at the farm as a business. At that time this was very dependent on account analysis and details of crop yields and animal performance figures. The era of gross-margin analysis and separating out fixed and variable costs had not yet arrived. This economics group numbered about twelve, and we would often set out in cars to inspect selected farms and co-operative farmers in Nottinghamshire and Lincolnshire. We would seek out as much physical data about the farms as possible and would be furnished with copies of recent farm accounts. We would then have to analyse the figures to determine which enterprises were contributing most to farm profits and whether there were any weak links in the system. The course put us in real touch with the farming world, a touch which had been lacking in those first two years of scientific study. That is not to criticise the science-based course because it was a springboard to specialisation over a wide field. For instance, in my particular year I can remember four different students doing honours degrees in animal physiology, agricultural botany, biochemistry and animal husbandry.

At last, the time for finals arrived. Lectures finished, there were no timetables except the timetable for exams which were spread over several weeks. Mostly we had to sit one exam per day, sometimes two. One of the requirements for taking finals was attending the exam wearing a university gown. I forgot mine for one exam and was not allowed into the exam room. I had to go back to my hostel to collect my gown. When I reached the exam room again all the examinees were scribbling furiously. I wondered what hope I had got for passing this exam, but I did pass. On the last day of the last exam, several of us went to the cinema in Loughborough to see the film *Carmen Jones*. It was light relief after the frenetic brain activity of the previous two weeks.

The results were announced before the end of the term. In our group of twelve we all passed for the BSc degree except one. He had failed, so we could not help feeling for him. No recognised qualification for him after three years of study. The highlight and culmination for the rest of us of our three years' study was attending the degree ceremony at the main university in Nottingham. My mother congratulated me on my achievement. Not so my father, although he did attend the degree ceremony.

Graduation, University of Nottingham, 1957.

Old farmhouse and buildings, Middlemarsh Farm, Sawtry.

The 'new' farmhouse, Middlemarsh Farm, Sawtry, built about 1917.

Threshing operation, Middlemarsh Farm, 1950s.

Cutting cereals with a binder, Middlemarsh Farm, 1950s.

Wedding, October 1964.

Our first property, Royston, South Yorkshire, 1965.

Our second home in Staincross, near Barnsley.

Our young family, Stephen and Susan, 1968.

Family photo, when Stephen was eighteen, 1982.

Our home in Somerset, 1987.

David at Pitminster Church, 1988. Licensed as a Reader in the Church of England.

Welcoming Princess Anne to the ADAS exhibit, Bath & West Show, 1987.

*Welcoming John Gummer to the ADAS exhibit,
Bath & West Dairy Show, 1987.*

Our home in Painswick, 1989.

Celebrating our silver wedding anniversary, in France, October 1989.

Shield for best small exhibit, Moreton Show, September 1991.

American friends' ranch-style home, San Antonio, Texas.

David with Uncle Len (£10 Pommie).

Our first granddaughter, Laura, Lewana, Western Australia, 1997.

Bungee jumping near Arrowtown, New Zealand.

Taiera Gorge Railway, Dunedin, New Zealand.

Church of the Good Shepherd, Lake Tekapo, New Zealand.

Ordination as priest by Bishop Michael, Gloucester, July 2004.

Golden wedding anniversary, October 2014.

FARMING

I was only twenty when I received my degree. That must have been something of a record. What to do now? I was still very keen on practical farming. Going back home to Middlemarsh Farm seemed to be the only practical route into farming. Obtaining a farming tenancy was impossible for a new graduate, and the capital required to buy a farm was prohibitive. There were plenty of opportunities in agricultural supporting organisations such as the agrochemical, fertiliser and feeding-stuff firms. There were also opportunities in agricultural education, consultancy and research. Some of my colleagues got jobs with county-based colleges called farm institutes. A few went into the colonial service, eventually being seconded to jobs in the African colonies such as Nigeria, Kenya and Rhodesia.

The Ministry of Agriculture had an advisory arm at the time called the National Agricultural Advisory Service (NAAS). This was established after the Second World War in 1948 to continue helping improve the productivity of British agriculture. It provided a free advisory service to farmers on technical and business matters and also to the supporting ancillary industries. This was another popular career avenue for agricultural graduates of my vintage. There were opportunities here not only in agricultural consultancy but also in research and development as the Ministry of Agriculture had established a string of experimental husbandry farms covering the country representing different soil types and farming conditions.

There were strong emotional pressures for me to return home to the family farm apart from my own inclination of wanting to farm. I mentioned in the last chapter that my father was struggling as a farmer, although he would have been the last to admit it. Crop yields on the farm were no better than average; the grassland was understocked and poorly utilised. Profitability was low, leaving little cash over each year for reinvestment.

My twin brother, Brian, had been working on the farm since he left school at sixteen. He had no agricultural qualifications and was therefore very dependent on my father. He knew that farm productivity could be improved, but he was not in a position to do anything about it. We had always got on well together and he looked forward to me coming home and providing a united front against my father. He had always looked to me for a lead anyway.

Perhaps the strongest influence on me was my mother, although I did not recognise it then. By the time I graduated her health was beginning to decline and in her heart of hearts I believe that she recognised that the future of the farm lay in my hands rather than her husband's. She had a dream of her two sons working harmoniously together, with me as the leader. She had bitter memories of a wartime partnership between her husband and his brother. They were always at loggerheads. The two families lived in adjacent farmhouses and there was continual conflict. This was at the background of her dream of her two boys working together. I think I offered her security for the future. She didn't feel secure with my father, who was always moaning about how difficult farming was and how he would like to retire. Retirement for him also was only a dream at his present level of means, and my mother realised that. She strongly urged me to return home and gradually take up the reins.

As a family we discussed my future, although my father was always non-committal. He could not cope with any challenge to his authority, and if challenged he got aggressive. However, we did hammer out some sort of deal. He agreed that we would farm in a family partnership when I was twenty-one. Obviously he would supply the initial capital for the partnership as Brian and myself had none. Profits from the partnership would be divided on a fifty-fifty basis with Father taking fifty per cent and Brian and me twenty-five per cent each. That at least gave me some scope for us to build up capital in the partnership, provided that we could improve the profits. The other valuable asset which Brian and I were contributing to the partnership was our labour.

After graduating, I arrived back home at the farm in July 1958. Now to start work! Not that work was new for me. I had spent every summer holiday helping to get the harvest in for as long as I could remember. I had also worked hard at college to obtain my degree. Middlemarsh Farm was situated between Sawtry and Glatton in what is now Cambridgeshire. When I was there it was Huntingdonshire. The farm was bounded by the Sawtry–Glatton road to the west and the A1 Great North Road to the east. The farmstead was on the Sawtry–Glatton road about a mile from each village. The buildings included an old barn and a relatively new Dutch barn. There was an implement shed and three crew yards for housing cattle in the winter. A cowshed with standings for about twenty cows was obsolete. A spacious stackyard adjoined the buildings on the north side.

The farm size was 280 acres, although about thirty acres was sublet to my uncle. The soil type was non-calcareous boulder clay, a difficult soil to work especially in wet conditions. Some of the fields contained basins filled with a lighter silty soil which had a much higher lime requirement than the surrounding clay. When I arrived on the farm the cropping was about fifty per cent arable and fifty per cent pasture. Crops included wheat, barley, oats and peas. My father also had a contract to grow fifteen acres of sugar beet for sugar. After the sugar had been extracted at the factory there was an opportunity to take some of the dried pulp back for animal feed. The feeding value of this was similar to oats. The grassland was stocked with cattle for producing beef. Some of the cattle were reared from bought-in calves; others were bought as stores for fattening.

During the months before my twenty-first birthday we arranged the details of our partnership agreement. Much of the detail was drawn up with my father's accountant. He was a funny little man, very formal and pedantic. I remember that there was one detail on which he was very adamant. The agreement was between a father and his two sons. If wives should arrive on the scene at a later stage they were not part of the agreement. The accountant warned that he had seen too many partnerships wrecked by wives. Sadly the good advice was later to be ignored by one of the partners. A failing of the partnership agreement was that it did not spell out the roles of the three partners. The only point which my father insisted on was that he signed the cheques. I think that he believed that if he signed the cheques it showed that he was in charge. But in fact he wasn't in charge because he never spelt out the roles of Brian and myself, and this became a source of friction in the future. However, straight away I took charge of the bookkeeping and the accounts, which gave me a good handle into the business.

I realised that I was not going to change the business overnight. My father could not have coped with dramatic change and would have become impossible. It was a question of gradually improving productivity through increasing cereal yields and better use of grassland. There was not enough capital available to materially increase livestock numbers, so the only practical solution was to reduce the area of pasture the livestock was kept on. This necessitated ploughing up some of the permanent pasture on the farm. Much of this was in ridge-and-furrow grassland, a survival of the nineteenth century. Ploughing out this pasture required permission from the estate, which initially my father was reluctant to ask for. However, after consistent badgering from Brian and myself he relented. I think he could see the logic of the argument, though he did not relish making the request. Anyway the estate agreed even if they were not overenthusiastic. Over the next two years we ploughed out about eighty acres of permanent pasture for cereal cropping. A sizable proportion of this attracted a grant of £12 per acre, so there was active encouragement from the government for this operation. Nowadays the wheel has turned full circle and ploughing up old permanent pasture would be prohibited. But remember the accent in the 1950s and 1960s was on increasing the productivity of British agriculture.

When I started farming at Middlemarsh much of the farm was rather poorly drained. This was limiting crop yields as well as restricting the timeliness of cultivation. Generous grants were available at this time supplying up to sixty per cent of the cost of approved drainage schemes. This was one area where my father did not need any convincing. He knew that sound drainage was essential and had always said that you cannot farm against water. Quite why he hadn't tackled it earlier I don't know. Perhaps it was shortage of capital or else he was lacking the confidence to set about the problem. Anyway, during the course of the next two or three years we arranged several extensive drainage schemes. Most of these included laying strategic tile drains and mole draining over them. The most ambitious scheme was digging out a main ditch right through the farm with a dragline. This crossed an eighteen-acre field and unearthed a previous drainage system which had perhaps been installed up to 100 years earlier. It consisted of horseshoe tiles laid on flat tiles. The amazing result was that after the ditch had been excavated many of these old land tiles began to run again into the new ditch.

One secret for improving crop yields was judicious use of fertiliser. Fertiliser firms such as ICI and Fisons expanded greatly after the Second World War. They manufactured a variety of fertilisers to suit various soil types and crops. My father bowed to my greater knowledge in this area, so we gradually increased fertiliser use and timeliness. Before I came on the scene he had used very little fertiliser. Much of the cereal crop became combine-drilled, which meant sowing fertiliser into the soil at the same time as the seed. This slowed down the sowing operation, but at least it was all completed in one pass of the machinery.

Another area in its infancy was crop spraying. This was another area where my father was amenable to my more up-to-date knowledge. We purchased our first crop sprayer and began using hormone sprays such as MCPA and mecoprop to control weeds. Weeds are a serious threat to crop yields, and to be able to limit their competition at a critical stage of crop development was a useful new tool in the husbandry armoury. Besides being able to control dicotyledon weeds, sprays were being developed at the time to control monocotyledons such as wild oats and blackgrass, which were becoming a serious threat to cereal yields in East Anglia.

When I started farming with the family the farm was not well mechanised. Harvesting was by the old-fashioned method of cutting the corn with a binder which converted a standing crop into a field of sheaves. These were gathered together into stooks, which later were carted to the stackyard and built into corn ricks. The ricks had to be thatched to protect them against wet weather. Then during the winter and early spring months the corn would be threshed out for sale. The threshing tackle was supplied by a local agricultural contractor. The straw was retained on the farm, mostly for animal bedding. There was some nutrient value in oat and barley straw, so some of this was eaten by the cattle. It does not take much imagination to see that the whole process of cutting the corn, stacking it and later threshing it was a long and laborious operation. Harvesting itself occupied up to six weeks. During my second year on the farm we purchased our first combine harvester, which completed all the harvesting in one operation. We bought the combine second-hand from an agricultural-machinery merchant together with a pick-up baler for baling the straw. This equipment totally transformed the harvesting operation and was much less labour-intensive. It did create another problem though. It was a bagger combine, which meant all the corn was delivered in sacks. The sacks we used were four-bushel railway sacks. Four bushels of wheat weigh two and a quarter hundredweight; barley weighs two hundredweight and oats one and a half hundredweight. You can imagine that it was extremely hard work loading these sacks of corn onto a trailer, then transporting them to the farm buildings. These buildings were ill-equipped to store grain in, and access to some was very awkward. However, overall it was a big step forward from binder to combine. By the early 1960s we had progressed enough to buy a new Massey Ferguson tanker combine, which was a further step forward.

Another area for attention was tractor power. My father had relied on a small Caterpillar tractor, two old Fordsons, and a small grey Ferguson. All the heavy cultivation was done by the Caterpillar, which became too slow to deal with the extra acreage of cereals. Initially we purchased a more high-powered Ferguson 35 together with some of the necessary hydraulic equipment to go with it. This helped considerably, not only with coping with the autumn ploughing but also with cultivations. After a few years we purchased an even higher-powered Ferguson 65, which totally transformed our ability to deal with cultivations at the right time. The whole operation was speeded up. The Ferguson 65 was also used after harvest to 'bust up' some of the cereal fields when conditions were dry enough. The busting operation meant pulling a tine through the soil at about eighteen inches deep to promote cracking of the soil, thus improving soil structure and drainage.

My father had used a one-year grass ley as a break-up crop. This was usually cut for hay, which was sold if surplus to the cattle requirements. I did not think a one-year ley was a very efficient break crop and introduced a three-year grass/clover ley in its place. This gave cereal land a longer break from continuous cropping and could also be utilised as grazing for the beef. We used modern grass varieties in these new leys, which were much more productive than the old permanent pasture we had ploughed up. Those consisted of unimproved grasses such as *Agrostis*, which were not very productive and were late starting spring growth. The new leys thus helped to compensate for the reduced area of grassland due to the ploughing-up campaign.

With the beef we tried to adopt an eighteen-month beef system. This meant buying in calves for rearing in the autumn with the aim of selling them out of yard a year the following spring. In practice it was difficult to achieve the desired results. Cattle only had one season of grazing from the age of about six months to a year. They were relatively young to make maximum use of grazed grass, so growth rates at grass were disappointing. Also they needed to be stocked fairly densely at grass for the system to be efficient. Feeding during the second winter had to be at a fairly high level to achieve the live weight gain necessary for the cattle to be fat enough for marketing before the next grazing season. We also ran into problems with bovine tuberculosis, which necessitated slaughtering some calves before they were six months old. To keep grazing stock rates at reasonable levels we had to buy in store cattle in the spring to replenish the calves lost. I began to have doubts about the profitability of the cattle enterprise as a separate entity, although it did complement the cereals. My father was rather swayed by the value of the cheques we received for the beef, but I was concerned about the level of input going into the enterprise, especially capital.

There were some diversions away from farming. When we were twenty-one Brian and I bought a new car, a Morris Minor 1000, with proceeds from the sale of four bullocks. This meant that we became more independent from Father. He had let us use his car when necessary, albeit with some reluctance. The drawback of a shared car was that we either had to go out together or agree when we could use the car separately. This was reasonably amicable until we started courting!

The family were all members of the Methodist Church in Sawtry. Fairly soon after I went back to the farm I started training as a Methodist local preacher. This entailed a study course on the Old and New Testaments and Christian doctrine.

I had to pass examinations in all three of these subjects set by the Local Preachers Department of the Methodist Church. The supervisor for my course was the local Methodist minister stationed at Huntingdon. This course of study and examinations took about two years, after which I faced an oral examination before the Local Preachers meeting of the Methodist Circuit. I was successful in all of these requirements and became an accredited Local Preacher in the Hunts Mission Circuit in 1959. After this I conducted services on most Sundays in the villages and towns of the circuit, which served most of the county of Huntingdonshire.

The Methodist minister from Huntingdon became conscious that there were young people like Brian and myself who were quite isolated in the villages surrounding Huntingdon. He arranged a youth fellowship meeting on Sunday evenings at his Huntingdon Manse where these young people could meet each other and also young people from Huntingdon Methodist Church itself. This became a good social opportunity for meeting members of the opposite sex as well as a fellowship group for learning more about the Christian faith. Brian soon began courting a girl called Joan and wanting more of the car himself! In the early years I did not form any deep attachment, although I became friendly with several of the girls from the group – particularly a girl called Jean. I also enjoyed brief friendships through meeting girls on Methodist Guild Holidays.

Brian married Joan in 1961 at Huntingdon Methodist Church. The old farmhouse at Middlemarsh had become vacant, so that became their home. Previously it had been used to house a farmworker. Brian's marriage slightly altered the dynamics of the partnership. Previously we had all met at breakfast time, when we could discuss the work programme of the day. Naturally Brian was no longer there for these discussions, so he began to feel left out. I don't think it made any difference to the decisions because they were made either by my father or myself, but at least Brian was there so I guess he felt he contributed. Joan fairly soon discovered that most of the decisions were being made by me and this rocked her security. There were obviously other discussions going on in the old farmhouse which my father and I were not party to. I found after a period of time that I could no longer count on Brian's support when there was a disagreement with my father. He began to hedge his bets and became much more awkward to work with. He started to question my decisions and judgements, not because they were wrong but because of Joan's influence. This takes me back to the partnership agreement, which was lacking in not defining the respective roles of the partners. Since the agreement had been drawn up I had become the manager in everything but name with the tacit acceptance of both my father and Brian. This obviously irked Joan considerably. The seeds of future conflict had been sown. Brian was also ignoring the accountant's advice to keep wives out of the business arrangement.

The matter came to a head as I began to contemplate my future. My friendship with Jean had blossomed into romance and we became engaged early in 1963. What future was there at Middlemarsh? I began to have doubts about my future there. The family conflicts were making me very unhappy, feeling lonely and isolated. At the back of my mind, I also thought that perhaps God was calling me to a richer, freer and more independent life elsewhere.

I had been instrumental in raising the technical efficiency of the business and improving farm profit considerably. Cereal yields had increased from an average of twenty hundredweight per acre to thirty-six hundredweight per acre, which is an eighty per cent increase. Profits had risen from practically nothing to over £3,000 per year, which was well above average farm income at the time. About this time I did a gross margin analysis of the farm business. This was a new method of farm accounting which helped dissect out the profitability of the individual farm enterprises based on separating fixed and variable costs. The fixed costs, like labour and machinery, serviced the whole business; the variable costs could be allocated to individual enterprises. My analysis showed that winter wheat was the most profitable cereal enterprise and oats the least profitable. Gross margins of sugar beet and beef were respectable, but were probably absorbing more than their fair share of fixed costs. This information was extremely useful for planning the future direction of the business, but who was going to make the decisions?

I realised that the future roles of the partners had to be more clearly defined. In my own mind my father should think of retiring. He was approaching sixty-five. I had some sympathy with Brian's position and felt he needed a more clearly defined role and a specific area of responsibility. With some difficulty I managed to get the partners round the table to discuss the future.

With my father the idea of retiring had receded. His life had become much easier and he was enjoying being seen in the farming community as successful. Neighbours looking over the farm hedge could see the visible improvement in cereal crops. Brian was very negative in these discussions. He was not prepared to agree to anything which he thought would upset Joan. For myself I wanted it clearly established that I would be the manager in the future and would be making the business decisions. It was clear to me that the time had come to make major changes in the enterprises to be retained and their future scale. Neither of the other two partners would agree to the changes I was asking for. They did not see why the present ill-defined situation could not continue. It was quite obvious to me that whereas I was prepared to put my cards on the table as to what I wanted they were not. My father still wanted it to appear as though he was the boss even though I was doing the bulk of the managing. Brian could not lose face with Joan by agreeing to me being identified as the manager. As far as Joan was concerned Brian was an equal partner with me in spite of his more limited ability, knowledge and drive. So there was an impasse. Without these changes I could not see a future for myself and a new wife at Middlemarsh in spite of my love of practical farming.

I decided to resign from the partnership. In this decision I was loyally supported by Jean, whose future was obviously bound up with mine. My mother was distraught; my father was bitter. Brian showed no emotion at all, but I guess Joan was very pleased. The decision to leave the farm was made more difficult by my mother. She had been diagnosed with breast cancer. What with that and the shattering of her dream of her two sons working harmoniously together the future looked bleak for her. She pleaded with me to stay but my mind was made up. I had to think of my own future. I felt guilty, but I was not prepared to carry on without a new formal agreement defining our respective roles. My mother stayed loyal to my father although I believe that she understood my position. She was too diplomatic to be seen taking sides. She also understood that Brian was being ruled by Joan, and my father too was smitten by her. This clouded his judgement in seeing the problem. My leaving the farm business provided a vacuum for Brian to fill without reference to me. Would he rise to the challenge?

AGRICULTURAL CONSULTANCY

I made my decision to leave Middlemarsh about Easter 1964. I was required to give three months' notice, so I planned to leave the farm after harvest. Jean and I decided to get married that autumn during October. Jean was doing a nursing SRN course at the time based in Nottingham. If all went well she would obtain her qualification at about the same time. So I had to get a new job before then.

It seemed logical to try and get a job in agricultural consultancy. I had the knowledge and qualifications and also the bonus of actual farming experience. I mentioned NAAS before, the advisory arm of the Ministry of Agriculture, Fisheries and Food (MAFF). There were also consultancy opportunities in the agrochemical, fertiliser and feeding-stuffs industries. Most of the major firms had developed their own advisory services to supplement their sales divisions. I thought that I would prefer to work in NAAS, which was independent and also covered the whole agricultural field, so I made my application. I was fortunate because NAAS was recruiting that spring and was interested in appointing graduates with experience as well as newly qualified people.

I had to go to headquarters in London for an interview, which I felt went quite well. I was quizzed particularly about the enterprises at Middlemarsh and the justification for keeping them. As I had done a gross margin analysis recently I was able to quote chapter and verse. This showed that I understood the business as well as being up to date on new methods of business analysis. A few weeks later I was notified that my interview had been successful and I would be appointed as a district officer in Yorkshire based on Harrogate, where the divisional office was situated.

A few words of explanation are necessary here to explain how NAAS operated. The organisation covered England and Wales but not Scotland. The country was divided into regions and divisions. The regions were managed by a regional director and housed all the specialist scientific services. These services included soil science, nutrition chemistry, plant pathology and entomology. There were also specialist departments in animal husbandry, agronomy, and farm business management. All of these specialist services were there to support the front-line advisory work in the divisions. Each division was roughly equivalent to a county, except some of the smaller counties were amalgamated together. Divisions were managed by divisional agricultural officers. Each division was divided into a number of areas, often supported by an area office. District officers worked from these area and divisional offices and were responsible for all agricultural advice in their allocated area. They were free to invite specialists from the regional office to help with problems which were outside their own expertise. The regional offices provided analytical services for soils, forage, pests and diseases. All in all, NAAS provided a very comprehensive free advisory service to the farming industry.

I was appointed to take up my post in early September 1964. Before then I had received a letter from the divisional agricultural officer at Harrogate which said I would be stationed at the area office in Barnsley. This was a bit of a shock to say the least after my initial expectation of working from Harrogate. The district I would be looking after would be based in Penistone, a small town at the southern end of the Pennines just north of Sheffield. What a change from East Anglia!

I remember clearly my first journey to Barnsley with my

Morris Minor 1000 stacked with virtually all my possessions. I had bought out Brian's share in the car when he got married. We had finished harvesting before I left the farm, but when I got to the Penistone district what harvest there was had still to be gathered. The grass was still very green, whereas grass in East Anglia was looking brown by then. This sent me an immediate message. This was grass-growing country, not corn country. That was a very important first lesson.

I had been fortunate in obtaining lodgings in Barnsley for a few weeks through the auspices of the Methodist Church. This gave me a little time to look around for a house where we could start our married life. Through a local advertisement I found a three-bedroomed terraced house in Barnsley to rent. Jean managed to come up to Barnsley for a weekend before our wedding to see our proposed starting home. She wasn't impressed by the black smuts everywhere. Barnsley was still a coal-mining town.

I had a lot to learn during my early years in NAAS. Although I was technically qualified I knew next to nothing about the internal workings of the organisation. Newly qualified graduates were given a two-year training period before they were allocated a district to manage. So I was thrown in at the deep end without this training period. However, I had the satisfaction of being paid more money than colleagues on the training grade. I had to learn about all the specialist departments at the regional office in Leeds and how to use them to provide a proficient service for my district. I also had to learn how to make maximum use of the office support staff at the area office in Barnsley. These included clerical and typing services.

One of my first actions was to apply for some leave so that we could get married. My new colleagues thought this was a bit of a joke as I was only just starting work with the organisation. The rules allowed for five days' special leave for marriage, which meant we could have a week's honeymoon. We got married at Huntingdon Methodist Church on 24 October 1964 and spent our honeymoon in Eastbourne.

The area around Penistone was mainly grassland. Farms were quite small by East Anglian standards. The average family farm was less than 100 acres. The main enterprise was dairying, with sheep being important on the upland and moors. Some dairy farms had developed their own retail milk rounds which virtually doubled the size of their business. Other supporting enterprises included beef, pigs and poultry. The Ministry of Agriculture was running a Small Farmer Scheme to improve the viability of small farms. This involved ploughing out permanent pasture and introducing grass leys to improve agricultural productivity. So ploughing out permanent pasture was still official government policy even in the late 1960s. This scheme was supervised by district agricultural officers, so I had to get to grips with this as well. It became rather an administrative nightmare because farmers kept changing their minds about their cropping. This was allowed, but only with an approved variation to their scheme. You can imagine that these small farmers in a Pennine district were not accustomed to looking five years ahead for their programme. That was the time limit for the duration of each individual scheme.

After six months in our rented accommodation in Barnsley, Jean and I bought our first property in Royston. It was a two-bedroomed bungalow situated in a cul-de-sac with about half an acre of rough ground at the back of it. I had the satisfaction of developing this into a garden over the next few years. It was a rather stony infertile plot. The better part I used for a vegetable garden; the rest was sown to grass and I planted some fruit trees for an orchard. My wedding ring is buried in that piece of ground somewhere! Our two children were born while we lived in that bungalow: Stephen in 1966 and Susan in 1968. Sadly my mother died of cancer in 1967, so she never lived to see Susan. She had always wanted a granddaughter.

In the early 1970s we decided to move from our bungalow as it was too small for our two children and entertaining guests. My mother-in-law visited us fairly frequently as she was a widow living on her own in Peterborough. There wasn't a lot of property to look at, but eventually we settled on a three-bedroomed bungalow at Staincross, situated between Barnsley and Wakefield. This had a developed ornamental and vegetable garden although the plot was much smaller than we had enjoyed at Royston. From the lounge we had extensive views over the Yorkshire countryside as far as the cooling towers at Ferrybridge. This was to be our home until 1976.

About this time the Ministry of Agriculture decided to close the Barnsley and Huddersfield offices and amalgamate them into one office at Mirfield between Huddersfield and Dewsbury. The change of office meant a much longer drive to the office in a direction away from my district. The ministry compensated me for this by allowing me to work from home, so I only went to the office on about two days per week. By now I had become very proficient in the farming systems of my district and had my own priorities for advice and development.

I had become very knowledgeable on dairying systems, grassland management and silage making. We introduced a silage competition, which was keenly contested among the leading dairy farmers of the district. It was judged by a member of our Nutrition Chemistry Department in Leeds supported by a chemical analysis. The competition was supported by the South Yorkshire Grassland Society, of which I had become a committee member. They awarded a cup for the winner, which was presented at their annual dinner dance. This was a major social event in the farming calendar. A lot of prestige hung on being the winner of the Silage Cup.

When I arrived in Penistone District most of the cows were housed in cowsheds during the winter months. These were demanding on labour and were restrictive when farmers wanted to increase the size of their dairy herds. The solution to this problem was to convert the cowshed into milking premises and house the cattle elsewhere. I became very involved in helping farmers convert from cowshed housing and milking to parlour milking, cubicle housing and self-feed silage. This enabled farmers to change to the new system at a fraction of the cost of building new cowsheds which had to conform with stringent Milk and Dairy Regulations. Under the new system only the parlour and dairy had to comply with these regulations. Also milk was pumped direct to a bulk tank, which was another big step forward in the collection, hygiene and transport of milk.

My involvement in these changes was to help the farmer in designing new premises. I would then draw up a budget to show the cost and future predictions of profit. The question always was, could the farmer afford to finance the changes and would it be financially worthwhile? The answer inevitably was that the new system would soon pay for itself, based mainly on the increased number of cows which could be kept. But the increased number of cows demanded higher stocking rates, improved grassland management and an adequate supply of good-quality silage for the winter.

Once a few farmers had converted to loose housing and parlour milking the uptake of this change began to snowball. I managed a number of major demonstrations using cooperative farmers who had made the change. Usually these farmers were so delighted with their new system that they were only too pleased to allow their farms to be used for demonstration purposes. Farmers flocked to these demonstrations and I tried to arrange for a major demonstration each year. These demonstrations were supplemented locally by meetings on grassland management and silage making. I took it as a major compliment that many farmers said that the name of David Newell was written all over their new dairying facilities.

My success as a district agricultural officer was noticed by NAAS senior management and I was promoted to senior agricultural advisor early in 1976. This meant a move to take charge of an area. It was always a bit of a lottery where the vacancies would be. Eventually I was allocated a post at Leamington Spa Area Office with responsibility for North Warwickshire. Two district officers would report to me and I would also manage a small district of my own. This was a first step onto the management ladder. I took up my new post in April 1976, the beginning of the hot summer. From April to September I commuted from Barnsley, staying in lodgings in Leamington Spa during the week. We moved to a five-bedroomed house in Leamington Spa early in September. I remember that the Halifax Building Society in Barnsley was appalled when we asked for a mortgage of £10,000. The same society in Leamington Spa granted the same figure without batting an eyelid.

My stay in Leamington Spa proved to be short-lived for a number of reasons. Looking back I think this was because of a weak divisional agricultural officer based at the divisional office in Worcester. He insisted on me taking responsibility for socio-economic advice in Warwickshire. This area of advice was to assist farmers wanting to diversify their business in enterprises outside agriculture. This wasn't an interest of mine and I resented being required to specialise in this area. After a time I also found that the divisional agricultural officer betrayed confidences I had shared with him about certain members of staff. I felt this was unforgivable in a manager.

However, there were some successes in Leamington Spa. Before I left Yorkshire I had applied to do a study tour in Holland on intensive dairy farming. This arose because of my work on intensive dairy farming in Peniston District and also my contacts with the South Yorkshire Grassland Society. The study tour was approved for June 1976. During the tour I visited several research organisations in Holland as well as some leading farms and farmers. Several farmers were suffering badly from the dry summer. Grazing grass was in short supply and many reseeded pastures had failed. My study tour entitled 'High Dairy Cow Performance from Grass and Silage' was written up, published and circulated within NAAS and the agricultural industry.

My expertise in dairying and grassland management was accepted by my colleagues at Leamington Spa and used in our promotional programme. I became active in the Warwickshire Grassland Society and introduced a silage competition along similar lines to the competition in Yorkshire. I spoke at farmers' meetings on grassland management and I was asked to judge a grassland competition in South Warwickshire. Also I organised a farm demonstration on grassland management.

One extra responsibility that I did welcome was taking part in a farming course for bank managers. This was based on the National Agriculture Centre at Stoneleigh. My involvement was to find a local farm planning a major new development. This ranged from buying extra land to planning a new dairy unit. Aided by our business management section we costed the development and made forecasts of future farm profit, including servicing the loan required from the bank. The bank managers were taken to the farm and given a copy of the budgets. On the strength of what they saw they then had to decide whether they would lend the money or not. On one occasion I remember because of a conversation overheard in a bar one bank seized a major project from a rival bank financing a cooperative grain store in South Warwickshire. Such is the way of business.

Another happy memory of Warwickshire was my association with Kenilworth Agricultural Discussion Group. I arranged for them to take part in a national farm business management competition. This involved farming a hypothetical farm and feeding our proposals into a computer system to predict profitability. Each month the parameters would change, so the farm plan had to be changed accordingly. We did not win the competition, but it taught the participating farmers a lot about farm business management.

My departure from Warwickshire was rather sudden and unexpected. After an annual appraisal I asked for an interview with the regional manager. Rather to my surprise he recommended a sideways move to Essex. He had worked in Essex early in his career and had retained happy memories of the county. It was a promotion in terms of technical demand and farm business understanding. We agonised over the decision within the family, but eventually decided to move to Essex. The attraction for me was that it was a demanding agricultural consultancy job rather than one where I was being asked to specialise in socio-economic advice.

The move to Essex did not get off to a good start. I started the job, based at the divisional office in Chelmsford, in early February 1979. It was a bad winter that year with persistent frost and snow, so commuting was difficult. A friend from Leamington owned a flat in Epping which was vacant, so I was fortunate to have the opportunity to rent that. It was very cold returning after weekends in Leamington, but at least it gave me independence. Finding a house for the family in Essex was a nightmare. House prices were rising fast and demand was high. It was also a problem sorting out favourable schools for continuing our children's education. By this time they had reached secondary-education stage. Eventually we settled on a new house in Great Dunmow on a small development of five houses. It proved to be an idyllic spot. The children were able to walk across the fields to the comprehensive school called Helena Romanes. It had a good local reputation. We moved there in June 1979.

Now to get to grips with Essex agriculture. My district was centred around Chelmsford with London clay soil to the south and chalky boulder clay to the north. It was an area of large farms and some very big farming companies. The average family farm was about 500 acres; some of the companies were farming several thousand acres each. What a contrast with the small Pennine farms I had worked with in Yorkshire. The challenge for every agricultural adviser is to become a specialist in the farming system of his district. In Essex this meant specialising in the technical aspects of cereals and combinable break crops, but also having a keen grasp of farm business management. In Essex also advisers were being asked to specialise in one of the minor crops and act as a consultant to colleagues. I retained my grassland interest and soon covered the county for grassland advice. Most of the grassland was utilised by large dairy herds, particularly centred around Epping.

At this time the accent was on maximising cereal yields,

particularly winter wheat. Certain continental systems for growing wheat were being adopted and becoming fashionable. ADAS got involved in trials testing these new techniques alongside more traditional systems. These aroused a lot of publicity and interest and were publicised with some large-scale demonstrations. I was amazed how much trial work on crops was being supervised by ADAS in Eastern Region. It was a new area to me, so I had to learn fast and find sites for trials in my own district. A large conference on cereals was being held annually in Chelmsford attended by 200 to 300 farmers. I soon became involved with the working party organising this conference. Nationally famous names in the farming industry were invited to speak at this conference, which attained a prestige of its own.

The work of ADAS was all about communication and getting a technical message across to as many farmers as possible. The demand for technical advice in the spring was particularly high. The only way we could cope with this was to form cereal groups of about fifteen members and service these groups by walking the crops on a member's farm each week. This would deal with all the current issues and the decisions which needed to be made. The essence of good husbandry was timely application of sprays to control weeds, pests and fungal diseases supported by appropriate applications of fertiliser. We introduced a Telephone Information Service to support our technical advice. Farmers could dial in to find out the latest information on pests, diseases and fertiliser timings and receive early warnings – for instance, of when to expect potato blight. We also published a monthly farmers' bulletin mailed to all farms called *Essex & Herts Notes*. These bulletins contained technical information and results of field trials, for instance on cereal and oilseed-rape varieties. I was the editor of *Essex & Herts Notes* for a number of years – a demanding but satisfying job.

I mentioned that when I was farming at Middlemarsh we managed to push the average cereal yield up to thirty-six hundredweight per acre. While I was in Essex in the 1980s many farmers were averaging yields of over three tons per acre of winter wheat. This was due to higher-yielding varieties and a better understanding and application of timely inputs. The success of the cereal groups led me to form a grassland group in the Epping area. This was attended by all the major dairy farmers in the area. Meetings on members' farms were monthly rather than weekly. These meetings emphasised that grass was a crop to be farmed just like arable crops. We covered grazing systems, grass varieties and mixtures, the fertiliser regime and silage making. As with the cereal groups this grassland group became a very efficient medium for disseminating advice and information.

I regard my time in Essex as the pinnacle of my technical career. Besides being demanding technically it also afforded me the opportunity to get involved in the farm business management of some large agricultural businesses. While I was in Essex the Ministry of Agriculture were operating Farm and Horticulture Development Schemes, which gave grants to enable businesses to become more efficient in labour and capital use. The schemes were monitored by the local agricultural adviser, who also gave a farm business report to participants. Some of the schemes I administered were among the biggest in the country. Paradoxically although the schemes were introduced by the European Union to help smaller farm businesses, it was the larger businesses in East Anglia which gained the maximum advantage.

Early in 1986 I was invited for a promotion interview to Divisional Agricultural Officer (DAO). By that time I had some experience of managing a division as I had had to deputise for my DAO on several occasions due to his illness. There was also a gap of a few months after his retirement before a new DAO arrived in Essex. The interview went well and I was offered promotion. Three divisions were vacant – Maidstone in Kent, Worcester and Taunton. Which one would I be offered? I was offered Taunton, which is the one I would have chosen, so I was well satisfied. It meant a move, of course, so further family disruption. By this time Stephen had left school and was working in London. Susan was doing a business course at Braintree College. Susan's course would not finish until July 1987, which meant our home would have to remain in Essex for another nine months.

I became the Divisional Agricultural Officer at Taunton in September 1986. I was now a manager and no longer a front-line technical adviser, so this was quite a career change. Unfortunately for me the structure and objectives for ADAS were changing fast due to a change in government policy. Charges were being introduced for ADAS services instead of the free service provided since the 1948 Agriculture Act. As a divisional manager I was given a target revenue for ADAS services in Taunton Division, which comprised the counties of Somerset and Dorset. Then before I had been in the post for one year it was announced that the post of DAO would be abolished. The two ADAS management posts in the division (DAO and Divisional Surveyor) would be combined. Up to this time the divisional surveyor had managed the Land Service, consisting of surveyors, and the Drainage Service containing land drainage officers. I competed for the new post at Taunton, but was unsuccessful. This was clearly unfair as the other managers in the region had competed against each other; unknown to me I had competed against someone from headquarters. It appeared like a political move further reinforced for me when I found out that I had scored more highly in the interview than some of the other managers appointed in South West Region. Also Taunton Division had been as successful as any in raising revenue from services. I was offered a new regional post, designated regional agricultural adviser with a remit to develop ADAS chargeable services in the region. A sop was that I could work from Taunton Divisional Office rather than the regional office at Bristol. This was just as well because Jean and I had purchased a new house at Pitminster, near Taunton, by then.

The new post of regional agricultural officer did not last long. I was charged with developing cropping services and also editing a new ADAS publication for farmers who subscribed to ADAS to be mailed out monthly. Another responsibility was to liaise with agricultural colleges in the region to explore the scope for their future use of ADAS services. ADAS had always liaised with the agricultural colleges, but this new policy would change the nature of the relationship. Somerset was the venue for the Bath & West Show at Shepton Mallet. ADAS had mounted an exhibit at the show for many years. The Show Society had been progressive in mounting other major events during the year, such as a dairy show and a grassland demonstration. I was given the responsibility of representing ADAS on the Show Working Party, which organised these events. The remit again was to explore the possibilities for developing ADAS revenue.

In 1988, another major change for me. The post of divisional head of ADAS at Gloucester became vacant due to retirement. This was an equal post to the one I had missed out on at Taunton. This time I was successful, and I started the new job towards the end of October. But it meant another move just at the time when we were happily settled at Pitminster. Eventually we bought a new house at Painswick, at the southern end of the Cotswolds between Stroud and Cheltenham. We moved there at the beginning of April 1989. Changes in ADAS organisation were continuing. The service was being prepared for agency status as a first step towards full commercialisation. As a divisional manager I was asked to increase ADAS revenue yearly, at the same time as accepting cuts in the ADAS budget.

The structure of the division was a divisional office at Gloucester and area offices in the other two counties of the division, in Devizes in Wiltshire and Chipping Sodbury in Avon.

In conjunction with the middle managers at senior agricultural adviser level we worked out an annual plan of revenue targets. This meant an individual target for each adviser. A few advisers who had been used to giving free advice for all of their careers could not adjust to the new environment. They had to be offered early retirement.

To some extent the new regime made annual appraisal easier. Previously there were no accepted standards for measuring the effectiveness of an adviser. The whole system was very subjective, like workload carried, amount of development work undertaken in addition to advice and standing amongst colleagues. With the new system the adviser was judged according to their success in raising target revenue. Not all advisers could be measured in this way because some were still involved in giving free advice on environmental schemes and also advising the Ministry of Agriculture on technical issues.

My job at Gloucester lasted three years. By that time ADAS was ready for agency status. The new structure would abolish divisions altogether and create larger units to be called 'business centres'. Existing managers at regional and divisional level would have to compete for the new posts of business centre manager. I did not fancy one of these new posts nor another move as by this time I was fifty-five. I opted for early retirement, which was granted with a pension payable immediately. I had some satisfaction at the end of my ADAS career in that Gloucestershire Division was the most successful in South West Region for raising ADAS revenue. It had surpassed Taunton, which at the time I left there was the highest earner. So my ADAS career was over. During the final few years I enjoyed the opportunity of being a manager. I coped well with the responsibility and drew satisfaction for steering the ADAS team in Gloucester Division through some difficult and testing times.

RETIREMENT

When I joined the civil service in 1964, civil-service jobs were considered jobs for life. I guess the same could be said of many jobs then, like those in banking or teaching. I never expected to be offered early retirement at the age of fifty-five.

So what next? I was a highly experienced technical consultant as well as having a proven record in management. It seemed logical to look for another job in agriculture to take me up to sixty at least, the normal retirement age for civil servants. But agriculture was a declining industry. As well as cutbacks in ADAS, commercial firms involved in agriculture and horticulture, such as fertilisers, feeding stuffs and pesticides, were reducing staff as well. Up to this time most of the large commercial firms employed technical staff to support them in their development and also to advise farmers and growers in the use of their products. They were all reducing the numbers of their technical support staff. The same was happening in agricultural education. Applications for agricultural and horticulture courses provided by county farm institutes and colleges were declining. The colleges were having to diversify into other areas to try and sustain student numbers. So courses in equitation, the environment and even sport started creeping into college syllabuses. Colleges which did not adapt to these changes fairly soon ceased to exist. All this did not bode very well for finding another job in agriculture.

There were some jobs becoming available abroad. The Soviet Union was breaking up at this time. Soviet satellite states were declaring independence and abandoning their socialist systems. Even agriculture had been state-controlled. Now state control was giving way to private enterprise. But there was no supporting structure to help develop private enterprise. So these states began looking at the West to see how the agricultural industry was supported by government-backed research, development and advice. They decided to set up advisory services of their own and they looked to the West to advise them on what sort of structure would be appropriate to their circumstances. Several of my colleagues who had lost their posts here due to the reorganisation took up the challenge and signed up for eighteen-month or two-year contracts with countries such as Turkmenistan. Conditions were primitive by our standards and the work difficult and challenging. A former colleague invited me to join him on one of these contracts, but I did not fancy a bachelor life and poor working conditions even if the salary was attractive.

By chance I noticed an advertisement for the National Trust, advertising for an agricultural adviser. The post would have a national responsibility, but would be based from an office in Cirencester, which was convenient for me. The salary being offered was only about half of what my ADAS salary had been. Nevertheless there were other attractions. My ADAS pension was payable immediately and would be unaffected. So I applied for the post and was invited for an interview. That cleared the first hurdle. Initially I was impressed that the National Trust was advertising for an agriculturalist because the trust was dominated by land agents with surveyors' qualifications. It seemed that the trust was broadening its horizons. At the interview the chairman asked me if I would like to take my jacket off. I said, "Do you intend to make me sweat, then?" This created a laugh and lightened the atmosphere. The interview went quite smoothly, but I had no idea what sort of impression I had made nor how many other candidates were interviewed.

Unfortunately I was not offered the job. I learnt later that they had appointed another surveyor!

So what next? I expected I would need another job to make up the income I had lost from my ADAS salary. But I was quite fortunate in having contributed to several life-assurance policies to cover my mortgage. These began to mature with good bonuses which were not part of the mortgage cover. So if a £5,000 policy matured at £10,000, say, I only had to pay £5,000 to the mortgage provider and I could keep the rest. These policies were not only enough to pay off the mortgage, but also sufficient to cushion the reduction in my income for the next few years. This made me realise that I did not need a full-time job for income purposes, so I began to think that a part-time job would be acceptable. Unfortunately part-time jobs in agriculture were as difficult to find as full-time jobs. I did do a few local gardening jobs, which occupied my time and also provided some spending money.

I was a reader in the Church of England. Readers are qualified to conduct non-eucharistic services in the church and also assist the clergy with pastoral work. Our benefice consisted of two churches, one in Painswick and the other in Sheepscombe. I regularly took services of morning and evening prayer at these churches and occasionally preached and led intercessions at communion services. I had no real experience or qualification in pastoral work except my experience as an agricultural consultant. But helping people with personal problems was another issue.

A vicar who had retired from parish work was running courses in clinical theology in a nearby village. Clinical theology was developed by Dr Frank Lake in the 1960s. Its basis was understanding different personality types and their subsequent behavioural patterns. Such knowledge is helpful in interpreting why people react in particular ways and how they can be helped. I decided to enrol on one of these courses, which was a two-year commitment. The course was based around two specific books: *Clinical Theology* by Frank Lake and

The Growth of Personality by Gordon R. Lowe. Each course consisted of a group of about twenty members. We attended a weekly lecture followed by group discussion. At the end of the two-year course we obtained a Certificate in Clinical Theology. I found this additional qualification very useful personally in understanding myself. I am sure it has helped me in my approach to people with personal problems, not least in helping those recently bereaved.

Another skill I lacked was how to operate a computer. I had relied on computers at work, particularly in relation to monitoring budget expenditure and forecasting and monitoring income from ADAS sales. But the computers were ably operated by office staff. I only asked the questions; they supplied the answers. But at home I had no staff, so if I was going to use a computer I needed to learn how to operate one. I enrolled at Stroud Technical College, first on a typing course, then on a computer course. The advantage of learning how to type properly was that operating a computer was not limited to a two-finger operation. The computer course was based around Microsoft Works programs. I had no sooner finished the course in the year 2000 and obtained a certificate in computer operational skills when Microsoft Works was upgraded to Microsoft Word. There have been many changes since then of course. Such is the pace of change in computer technology! But at least I was qualified in using a computer. Being computer literate has been a great asset since, particularly with word processing, emails and the Internet.

One opportunity which retirement provides is being able to take longer holidays. We had always taken an annual holiday of two weeks with various breaks in between. Taking a longer break whilst at work was impracticable because of the build-up of my workload while I was away. The whole idea of a holiday is a break, refreshment and an opportunity to recharge one's batteries. There is no point in taking a longer break if that means returning to a mountain of work with the added pressure of what that entails.

Our first holiday after retirement was to America, which we had never visited before. We had some American friends who lived in San Antonio, in Texas. They had often invited us to visit them and now was an opportunity. We flew out to Texas in the autumn of 1992. Our American friends lived in a ranch-type house on the outskirts of San Antonio. It was on a large plot with plenty of space both inside and out. The rooms were spacious and well appointed. We were allotted a separate part of the house which was virtually self-contained, with our own bedroom and bathroom. The house had attached garaging for up to three cars, and an additional storage area. The large garden also contained a swimming pool. Although it was October the temperatures were high enough to use the pool comfortably. We were astounded at the comparatively low price for such a property compared with British prices. It was valued at about £70,000 in our money at the time!

Our American hosts set out to ensure that we saw all the local sights and enjoyed our stay. San Antonio is a lovely city with a river running through the centre, which always adds to the attractiveness of any town. We explored the city, we visited the Alamo and we went on the river. During our stay our American friends went off to a conference for about a week. We took the opportunity to go and see a friend from my childhood village in Huntingdonshire. She had married an American and lived in Arizona. We flew there on an internal flight. The planes for flights of this sort are on a tight schedule, being in and out of the airport in about half an hour. Jennifer was there to meet us at the airport at Phoenix to transport us to Mesa, where she lived. It was a happy reunion. I had not seen her since 1964, when I left the farm at Sawtry. We had both been members of the local Methodist church there and she had been the organist. She had kept her Methodist connections and was now a secretary at her local Methodist church in Mesa. While at Mesa we visited the Grand Canyon. Jennifer and her husband drove us there and back in a day, a distance of over 500 miles. We had a lovely day. The weather was perfect and the colours in the canyon magnificent. Jennifer, who had visited the canyon several times before, said she had never seen it look more colourful.

After returning to San Antonio, we used our friend's car to travel down to Corpus Christi. That was another unforgettable experience, driving a people carrier with automatic gears and on the right-hand side of the road. However, we soon got used to the wide carriageways, cruise control and long distances. Our friends were so generous and trusting, letting us use their vehicle with no previous experience of driving in America and automatic gears. Corpus Christi is another fascinating city and we were fortunate in finding a good hotel for our stay. It is situated on the coast of the Gulf of Mexico. Mentioning Mexico, we ventured into the country of Mexico with Nancy, our American hostess. She booked all three of us into a hotel on the border. We stayed in one room, which we were assured was normal American practice. It was new to us! The next day we walked from the hotel over the bridge spanning the Rio Grande into Neuvo Laredo, in Mexico. We spent a fascinating day there. Nancy was clearly nervous of our wandering off the main streets because of the security risk. It was clearly noticeable how poor Mexico is compared with its rich neighbour.

Over the next ten years we made three visits to Australia and New Zealand. I had an uncle and cousins living in Queensland. My uncle had emigrated to Australia after the Second World War as a £10 Pommie. After my retirement we also had a daughter, Susan, living in Western Australia. She had married an Englishman living in Australia at a place called Toodjay, situated about sixty miles north-west of Perth. We were not able to attend her wedding; but we visited as soon as possible afterwards, which was during the autumn of 1995. She and her husband, Mike, lived on a small farm unit with a self-built house. Conditions were very primitive by our standards. It is an arid area in the Avon Valley. Water was in short supply. They relied on rainwater collected in a tank off the roofs of the buildings. There was also a pond on the farm providing water for the animals. Whilst visiting them we took the opportunity of going to the coast to a place called Busselton. We stayed in a motel there, which was much more comfortable. Busselton is notable for having a mile-long pier popular with fishermen. We walked to the end of the pier, which health and safety would not have allowed in England. The pier was in a poor state of repair, with planks missing and no side rails. We did observe one fisherman catch a kingfish, which was large. They wheeled his catch off the pier in a pram.

From Perth we flew to Brisbane via Adelaide, a journey of five hours. This registered for us how big Australia is. Just think of where you could get to with a five-hour flight from Britain! We stayed a couple of nights with an Australian cousin in Brisbane before being transported to my Uncle Len's home at Gympie, some seventy miles north of Brisbane. He was living alone having lost his wife Marie to cancer just several months previously. We were impressed with his homestead, which he had had built for his retirement just outside Gympie. It was a spacious, ranch-style house with about an acre of garden growing exotic fruit like bananas and mangoes. We enjoyed our stay there, exploring the local countryside and also visiting two other cousins and their families who lived nearby. While we were there Uncle Len went down to Brisbane for a few days to celebrate a family birthday. We took the opportunity of driving north in his car, a four-wheel-drive Subaru to Bundaberg, which lies in a sugar-cane-growing area. From there we flew out to Lady Elliott Island in a small eight-seater plane, which landed on a grass airstrip on the island. The island is situated at the southern tip of the Great Barrier Reef. It is noted for its seabirds, and turtles which lay their eggs on the beach. Our accommodation for the night was a two-berth camping hut. It was quite comfortable, but sleep was elusive because of the continuous squawking of the seabirds. Our breakfast the next day was served outside a café area. A notice said, 'Don't feed the birds'; it might have continued, 'They will feed themselves.' A neighbouring table to ours was left unattended for a moment and was totally cleared of food in a flash, rashers of bacon and all. On the return journey to the mainland there was just the pilot and us. We sat with him in the cockpit, taking it all in. An unforgettable experience.

We stayed for three weeks with my uncle. He was disappointed that we did not stay longer. From Brisbane we flew to Sydney and stayed with a university friend for a few days. The highlight of our visit to her was meant to be a visit to the Blue Mountains. However, when we arrived there the mountains were covered in mist so all views were obscured. This was not an uncommon feature, we learnt afterwards. We were able to spend a couple of days in Canberra, the seat of the government in Australia. We actually sat in the parliament building while parliament was in session – something we had not done then at Westminster.

From Sydney we flew on to New Zealand via Melbourne, landing at Christchurch in the South Island. We spent three weeks in New Zealand: two weeks in the South Island and one week in the North Island. We hired a car from a firm in Christchurch which had been recommended by a friend. The car was a Honda Civic, which proved admirable for our purpose. Throughout our stay we used motels which were equipped to a very high standard. There are a number of competing chains, which keeps the price very competitive. We found all of them very economical, certainly much cheaper than guest houses or hotels. We found it relatively easy to prepare evening meals and eat in comfort, relying on local shops to provide all we needed in the way of groceries and vegetables. We would buy in enough to cover breakfast the next morning and a lunchtime picnic for the following day.

From Christchurch we drove down the eastern side of the country stopping firstly at Timaru then motoring on to Lake Tekapo and Mount Cook. Lovely features of the roadsides were dwarf lupins fully in flower and with a wide spectrum of colours. Driving in New Zealand is such fun. There is hardly any traffic on the deserted roads and you can stop almost anywhere. At Lake Tekapo we visited the Church of the Good Shepherd, which is sited with glorious views over the lake. It is noted for a cross on the altar table which is silhouetted against the view over the lake.

Our next stop was Dunedin, so named because many of the original people originated from Scotland. Dunedin has much to offer the tourist. It has a famous elegant railway station. A train up the famous Tieri Gorge is a must. When we were there the hillsides were covered in flowering yellow broom. Devon cream teas were served on board. Another must in Dunedin is a visit to the Albatross Centre situated at a windy outcrop at Otakou. We were fortunate to be able to see the impressive birds nesting and also view a video on their life cycle. From Dunedin we motored on to Invercargill, the most southerly town of the island. We visited Bluff, right at the tip of the South Island, from where there are views of Stewart Island, which we did not venture on to.

The next stop was Te Anau at the side of Lake Te Anau. This was a good spot from which to view Doubtful Sound, so named because Captain Cook thought it doubtful that he could manoeuvre his ship inside the sound. To get to Doubtful Sound is a boat crossing of Lake Manapouri, then a bus journey over the Southern Alps to the beginning of the sound. Crossing the alps we made a detour several miles into a mountain to see a vast hydroelectric scheme. This provides the power for Invercargill and the major steelworks nearby.

When we reached the sound it was raining, as it is most days. In this part of New Zealand and right along the west coast the rainfall is measured in metres rather than centimetres. The rain doesn't stop the boat trips on the sound, which run on 364 days of the year. Much of the west coast is not serviced by roads, so we had to motor inland to Queenstown, which is famous for water sports and bungee jumping. We didn't try bungee jumping or white-water rafting, but we saw some who did! The short route from Queenstown to Wanaka was on an unmetalled road, not recommended for hire cars, but we risked it anyway. It proved not too difficult. Wanaka is a lovely town, an ideal spot in the centre of the South Island and a good stopping point for several days to view the central area.

Continuing our journey northwards we made for the mountainous west coast. We stayed for one night at Franz Joseph, hoping to go and view the glacier. However, a key bridge had been swept away by floods and the road to the glacier was closed. Bridges are often swept away in this part of the country. We continued up the west coast as far as Greymouth. It is aptly named because after all the glorious scenery we had witnessed Greymouth has little to commend it.

We now needed to get back to the east coast to pick up the ferry from Picton to the North Island. We stayed overnight at Howard Springs, notable for its hot thermal baths. The motel we stayed in there was one of the best, superbly appointed. Swimming in the thermal baths was a real experience, but the water was too hot to stay in for long. Having arrived at Picton, we were fortunate to be able to take our car over to the North Island. Many of the car hire firms don't allow their cars to go over on the ferry. Our firm had a depot in Auckland where we could take the car at the end of our journey.

The ferry crosses from Picton to Wellington, which is the capital city. The North Island is more heavily populated than the South Island, but driving is still easy and pleasant. The climate is warmer. We visited Palmerston North and Napier. Napier is an art deco city, rebuilt after a violent earthquake in 1931. A stop at Rotorua then before journeying on to Auckland. Rotorua is famous for thermal springs, mudbaths and a lot of other thermal activity. It is like entering another world and the smell of sulphur is never far away.

We reached Auckland in time for Christmas, arriving on Christmas Eve. We had booked into the Kiwi International Hotel, which proved a major disappointment. We had booked a hotel so that we had somewhere for Christmas dinner on Christmas Day – or so we thought. When we arrived at the hotel we found that they were not serving meals on Christmas Day, except breakfast. However, I was fortunate in overhearing a conversation at reception where a local restaurant proprietor reported that he had two places unbooked for Christmas lunch. I immediately said I would take them. It proved an inspired reaction because it was an excellent restaurant even if expensive. Another bonus was that Auckland was deserted, so after our meal we took the opportunity of touring round in the car. So Christmas Day did not work out so badly after all.

Our final few days in New Zealand were spent motoring up to Cape Reinga, the northern tip of the North Island. This entailed going through a kauri forest and also travelling along Ninety Mile Beach. Unfortunately when we arrived at Cape Reinga it was shrouded in fog, so we were denied the views. We travelled back to Auckland to deliver the car back to the hire firm. Then the flight home via Los Angeles. It had been a memorable thirteen weeks and the holiday of a lifetime.

Space does not permit me to write up the other two visits to Australia and New Zealand. However, two experiences I will mention. We returned to Australia in the autumn of 1995. In the meantime our daughter and her husband had moved from Toodjay to a place called Lewana, well south of Perth and near a village called Ballingup. Since our first visit they had become a family, producing a daughter called Laura, our first grandchild. They had moved to a new job looking after a campsite which consisted of former woodsmen's huts sited in a park area. The accommodation provided was much better than their Toodjay house and much more spacious. Also water supply was not a problem. We stayed at Lewana for several weeks and were there for Christmas. The real purpose of our visit was to see Laura, and she was a delight.

On the further visit we were fortunate enough to travel on the Ghan Railway from Darwin to Adelaide the first year it was opened, in 2004. The journey took three days with a stop-off at Alice Springs. We travelled Red Star, which means we had a cabin to ourselves which converted to two bunk beds overnight. During the day we were free to wander along the train to the restaurant and seating area. The train made several stops along the way to view special features of the landscape.

We broke our journey for two days at Alice Springs. We could have stayed longer. It is a fascinating place with much to see. We could have visited Ayers Rock from there, but there wasn't sufficient time.

ANOTHER CAREER

I mentioned that after my retirement from ADAS I looked around for another job in agriculture. That search proved fruitless, but perhaps God had other plans for me. I was a Christian and an Anglican reader, but I had been thinking of a deeper commitment to the Church. This was encouraged by being a member of a local ministry team in Painswick whose purpose was to further lay ministry in the Church. This involved participating in a two-year training programme covering aspects of lay ministry like pastoral care, worship leadership, leading prayer, and bereavement visiting. The scheme was extended in the Gloucester diocese by providing opportunities for selected members of the local ministry scheme to train for ordination as non-stipendiary local ministers. My team encouraged me to put my name forward.

The process was to be interviewed by the director of ordinands and then by several lay interviewers appointed by the diocese. If all these interviews were positive there was a final interview with the Bishop. I graduated to an interview with the Bishop, which resulted in me attending a residential selection course. The course was quite rigorous, involving doing several written assessments, attending further interviews and doing various practical exercises like chairing small groups to tackle a particular set problem. The interviews covered intellectual capabilities, spiritual awareness and pastoral sensibilities. I emerged from this course with a glowing recommendation that I should attend a training course for ordination. This confirmed for me a call from God for deeper commitment and ministry in the Church. I began my course with the West of England and Midland Training Course (WEMTC), based at Gloucester. The normal course was of three years' duration, but this was reduced to two years for me in view of my age and experience. I began the course in the Autumn of 2001, when I was sixty-four.

The course involved attending an evening class once a week at Gloucester, participating in a residential weekend each term and writing up essay assignments on a regular basis. It was like being back at university again, which comes a bit hard when you are in your sixties. I guess it was good for the soul! The residential weekends were enjoyable. They were attended by students from Bristol, Gloucester, Worcester and Ludlow. About half of the students were women, which is an indication of the much greater proportion of women becoming ordained than in the past. Also they were all mature students. Younger students with a career in front of them tend to attend a college course full-time. The group represented a diverse range of ages, experience, ideas and churchmanship, which was interesting, challenging and thought-provoking. It was also testing on levels of tolerance.

The lectures varied in quality, as do university lectures. Subjects covered included theology, scripture, doctrine and pastoral studies. The essay assignments were related to the lectures, but also required additional reading. The lecturers marked the essays, which all counted towards suitability for ordination at the end of the course. My marks hovered around sixty per cent, which was well above the pass level. Another feature of the course was leading worship and also preaching, the congregation being the course members and also the staff. This was done along with assigned colleagues and was also marked. Working alongside fellow students with contrasting attitudes and ideas was quite a challenge, but I guess it was an indication of real-life situations and therefore helpful.

Before ordination there is a compulsory retreat to attend, which climaxes in an interview with the Bishop. Theoretically this confirms suitability or unsuitability for ordination, although normally unsuitability would have been detected well before then. So in reality it is a rubber-stamping exercise with hopefully some good words of advice and encouragement from the Bishop. My final interview was in June 2003 with Bishop David, the Bishop of Gloucester. I was ordained by him along with eleven others in Gloucester Cathedral on 29 June 2003 as a deacon in the Church of England. This was a moving and emotional occasion before a congregation of 900 people, including my family seated in reserved seats near the front of the cathedral. I was licensed to the Beacon Benefice, which included Painswick, Sheepscombe, Cranham, Pitchcombe, Edge and Harescombe with Brookthorpe along with Dr Helen Sammon, who was ordained at the same time. We were curates to support the vicar, the Reverend John Longuet-Higgins. What was encouraging was the support of many parishioners from these villages who attended the cathedral service and then entertained us to a reception at Painswick. So began my third career, as an ordained minister in the Church of England.

The role of the deacon in the Church of England is very much a servant role. The accent is on pastoral work, leading worship and preaching. The deacon is excluded from leading the sacraments of the Church, which include the Eucharist and weddings, and from proclaiming blessings. My appointment was as a non-stipendiary minister, which meant that I wasn't paid. I committed to doing three days per week and Sundays, with one Sunday off each month. Looking back I see that this was quite a commitment after retirement for eleven years!

One new experience for me was conducting funerals, both at church and at crematoriums. Strangely I found this a very fulfilling time. The opportunity of ministering to families at a particularly traumatic time in their lives is both an honour and a privilege. I appreciated the opportunity of bringing help, comfort and reassurance to the bereaved. An interesting aspect of this work is finding out about the lives of the deceased, which often reveals information about them which was not common knowledge. How often I have heard comments after a funeral like "I wish I had known that when they were alive." Conducting funerals means working closely alongside funeral directors, several of whom I have got to know very well and they have become personal friends.

Before becoming ordained, I was unaware of the amount of committee work the clergy were involved in. Automatically I became a member of the Parochial Church Council (PCC) the Deanery Synod and the Chapter of Ministers in the Deanery, all with a regular pattern of meetings. There was also a programme of continuing ministerial education (CME) devised by the Ministries Department of the diocese. This involved attending study days organised by the diocese and a residential weekend per term. There was also a requirement to attend a cathedral service on Maundy Thursday led by the Bishop, which involved a recommitment of ordination vows.

My year as a deacon passed very quickly and very soon I was attending another retreat in preparation for being priested. During the year Bishop David, who ordained me as a deacon, had retired and his place was taken by Bishop Michael. As with the previous ordination retreat, the climax was an interview with the Bishop. I was ordained as a priest in Gloucester Cathedral on Saturday 3 July 2004, among the first group that Bishop Michael ordained. Bishop Michael remained as the diocesan bishop for the next ten years.

The following day, Sunday, I presided at my first Eucharistic service at St Mary's Church, Painswick. There is something very special about one's first Eucharist, and I felt this at Painswick. It is a responsibility, an honour and a privilege, and I approached this Eucharist with apprehension but also excitement. It was also a thrill to perform my first Eucharist at Painswick, my home church, which had given me such encouragement during the years of preparation. That service in the morning was followed by a benefice party in the evening sponsored by Helen Sammon and myself in appreciation of the benefice support we had received. So began three years as curate in the Beacon Benefice for both of us. I followed my curacy with two years as associate priest in the Beacon Benefice. I retired when I was seventy-two, which is a requirement for non-stipendiary priests. Normally stipendiary priests have to retire by the age of seventy. However, I continue to be licensed as a priest, so I still conduct services and assist with pastoral work.

Apart from the Eucharist, the other services I have been taking as a priest are officiating at weddings and baptisms. Most of the weddings I have taken have been at Painswick, but I have also taken some at other churches in the benefice. A particular joy has been when wedding couples have returned a few years later as parents and I have had the privilege of baptising their children.

Retirement ministry is different because how much I do is very much dependent on my own inclinations and how much help the local vicar requires. In the Beacon Benefice the only appointed clergy are the vicar and one part-time non-stipendiary priest. With six parishes and six churches to serve this would be impossible without the help of retired priests. This benefice is not atypical, which illustrates how much the Church of England depends on retired clergy and non-stipendiary priests. Having said that, it is still a joy to be able to continue serving in the Church as a priest and pastor.

It might seem that there is a wide difference between working in agriculture and working in the Church. In fact there are many similarities, particularly between farming and the Church of England. Going back to the eighteenth and nineteenth centuries many vicars had links with the landed gentry and England was largely composed of rural communities. Tithes payable to the Church were common, paid either in cash or in kind to support the local incumbent priest. Where payable in kind the produce was stored in the local tithe barn. Some of these barns have survived to the present day and have become tourist attractions. Parsonages or vicarages were large and often had land attached to them. In my own lifetime one of my uncles bought a local vicarage along with the attached thirty acres of land.

The real affinity of the Church with agriculture is that they both work to a seasonal cycle and on some occasions during the year these cycles overlap. The Church has Advent, Christmas, Epiphany, Lent, Easter and Whitsun. Farming has ploughing, sowing, growing and harvesting. At various times these seasons come together. For instance, Plough Sunday used to be celebrated regularly on the first Sunday after Epiphany, which is 6 January. Traditionally when ploughing was done with horses and before tractors came on the scene ploughing occupied the major part of the autumn and winter months on arable farms. Early January was about halfway through the ploughing season and Plough Sunday was used to bless the ploughs and honour the ploughmen.

During the growing season it was customary to ask for God's blessing on the crops at what is called Rogationtide. Parishioners would walk around the parish boundary with their vicar, who would bless the crops as they walked. This custom was known as 'beating the bounds'. Usually there would be a service in the church either before or after the parish walk.

Another important shared occasion is harvest festival when thanks are given for the gathered harvest. The origin of harvest festival goes back to Old Testament times, although the celebration of a church harvest festival service was initiated more recently by the Reverend R. S. Hawker, Vicar of Morwenstow in Cornwall in the 1840s. An annual church celebration of harvest was first recognised officially in the Church of England in 1862.

Another link with the countryside is some of the parables of Jesus. He often used illustrations from everyday life in his teaching. Perhaps his best known parable is 'The Parable of the Sower', whom he would have seen scattering seed in a field.

Other nature parables include 'The Wheat and the Tares' and 'The Mustard Seed'.

Last year Jean and I celebrated our golden wedding. This coincided with us moving house to a bungalow in the village with a smaller garden. Our previous home, besides being a four-bedroomed house, also had a large terraced garden. As there were only the two of us now we decided to downsize, which seemed sensible anyway as we were not getting any younger. Unfortunately our children were not able to celebrate with us at the time of our anniversary, so we celebrated ourselves by going on a cruise to Europe. The ports of call were Antwerp, Amsterdam and Rouen, and we were able to do excursions from each of these destinations as well as exploring the cities themselves.

It seems unbelievable that we have been married for over fifty years. Those years have gone quickly, but have been happy, eventful, sometimes challenging but always rewarding. We have lived and worked in many different parts of the country, each with its own special memories. We have had opportunities to travel widely. Our children have produced families of their own, so we are blessed with four grandchildren: three girls and one boy, ranging from nine to seventeen. I have enjoyed three separate careers. I look back with satisfaction and forward with faith.